Green Energy and Technology

More information about this series at http://www.springer.com/series/8059

Vanessa Rauland · Peter Newman

Decarbonising Cities

Mainstreaming Low Carbon Urban Development

Springer

Vanessa Rauland
Curtin University Sustainability Policy (CUSP) Institute
Curtin University
Perth, WA
Australia

Peter Newman
Curtin University Sustainability Policy (CUSP) Institute
Curtin University
Perth, WA
Australia

ISSN 1865-3529 ISSN 1865-3537 (electronic)
Green Energy and Technology
ISBN 978-3-319-15505-0 ISBN 978-3-319-15506-7 (eBook)
DOI 10.1007/978-3-319-15506-7

Library of Congress Control Number: 2015932974

Springer Cham Heidelberg New York Dordrecht London

Printed on acid-free paper

Springer International Publishing AG Switzerland is part of Springer Science+Business Media (www.springer.com)

Foreword

We have known for decades now that a more sustainable planet requires significant attention to climate change and that reducing carbon output from human settlements is absolutely imperative for the future of human civilisation. However, most of the effort to monitor and assess carbon and other greenhouse gas emissions has been at the level of the individual building, and of those various and often competing schemes to assess and/or certify green buildings. The terrain is muddy at best. Rauland and Newman have done us a great service by dissecting this mess with tremendous clarity and in drawing our attention to the level of the precinct or neighbourhood as a better focus than the city itself, which on its own is a valuable insight.

Having established this perspective, Rauland and Newman offer an overview of several international and Australian eco-city case studies which they then evaluate in terms of their carbon claims, setting the stage nicely for the following detailed chapters on carbon neutrality and measurement, GHG accounting frameworks and rating tools for cities and precincts, and carbon-neutral certification schemes for the built environment. Their presentation is clear and helpful as there are many difficult issues and tangents associated with this material.

The clarity of analysis and presentation in these chapters is exceptional, and that in itself is a valuable addition to the literature. However, it is in moving forward from this foundation that the work becomes particularly interesting and where Rauland and Newman's contribution to knowledge becomes most apparent. They offer a framework for calculating the emissions associated with precinct-scale low-carbon development, and outline some core elements to be considered when establishing a certification process for such development. Their core elements include site preparation and construction processes, embodied carbon in materials, energy production and management, water management, waste management and travel, analysed before, during and after construction. This framework will not satisfy everyone interested in sustainable development, but it stays focused on carbon reduction and makes it measurable and manageable.

This carefully focused approach lays the groundwork for a final discussion of the benefits, barriers, opportunities and governance associated with low-carbon development. Specific barriers to the uptake of such low-carbon development are outlined, followed by specific opportunities for overcoming those barriers. Rauland and Newman demonstrate that this transition will not be easy, but nor is it impossible; indeed, most of the opportunities for overcoming these barriers exist already. Someone needs to connect the dots and package them together in a way that enables the low-carbon transition to proceed apace, and this book has done a better job than anyone else so far in doing exactly that.

Rauland and Newman's concluding chapter has some tantalising ideas for where the decarbonising of cities could go. In particular, they hold out a most promising idea, namely that enabling carbon credits to be generated from low-carbon developments and made tradable could provide a critical new dimension to climate change policy, making possible a mechanism for trading carbon within the built environment and creating a significant financial incentive for decarbonising cities. They suggest a possible way this could be approached and thus have laid the foundation for a global debate on how decarbonising cities can be mainstreamed.

Professor Mark Roseland Ph.D., MCIP, RPP
Director, Centre for Sustainable Community Development
Professor, School of Resource and Environmental Management
Faculty of Environment, Simon Fraser University
Vancouver, British Columbia, Canada

Acknowledgments

The research underpinning this book was undertaken during an Australian Research Council (ARC) funded research project, 'Decarbonising Cities and Regions'. We would therefore like to acknowledge and thank the ARC for their valuable contribution and support. We would also like to acknowledge the wider Decarbonising Cities and Regions project partners, including Horizon Power, Cedar Woods, Strzelecki Holdings and Parsons Brinckerhoff, as well as our colleagues at Murdoch University—Professor Goen Ho, Dr. Martin Anda, David Goodfield and Joanne Stewart. A big thank you goes to our project team and colleagues at CUSP, Dr. Colin Beattie and Dr. Jessica Bunning.

A number of other people provided valuable input and clarification on various specific subject areas, including Simon Dawkins; Jeb Brugmann; Matthew Warnken; Bruce Easton; Melanie Bainbridge; Meg McDonald; Professor Mark Roseland; and Dr. Samantha Hall. Thank you for your assistance. We are also very grateful to Dr. Robert Salter and Phillip Webster, who helped enormously with the editing process.

Finally, to our families, thank you for your continuous support and encouragement.

Contents

Abbreviations

ACCC	Australian Competition and Consumer Commission
ACCU	Australian Carbon Credit Unit
ASBEC	Australian Sustainable Built Environment Council
BAU	Business As Usual
BID	Business Improvement District
BSI	British Standards Institution
C&D	Construction and Demolition
CBD	Central Business District
CDP	Carbon Disclosure Project
CFI	Carbon Farming Initiative
CNP	Carbon Neutral Program
CNPG	Carbon Neutral Program Guidelines
CPM	Carbon Price Mechanism
CSIRO	The Commonwealth Scientific and Industrial Research Organisation
CSR	Corporate Social Responsibility
CUSP	Curtin University Sustainability Policy Institute
D&D	Demolition and Disposal
DAP	Direct Action Plan
DOIC	Domestic Offsets Integrity Committee
EE	Energy Efficiency
EMC	Environmental Management Committee
EmE	Embodied Energy
EMP	Emission Management Plan
ERAC	Emissions Reduction Assurance Committee
ERF	Emissions Reduction Fund
ESCo	Energy Service Company
GFC	Global Financial Crisis
GHG	Greenhouse Gas
GRID	Greening, Regeneration and Improvement District

ICLEI	Local Governments for Sustainability (Formerly 'International Council for Local Environmental Initiatives')
IPCC	Intergovernmental Panel on Climate Change
ISO	International Organisation of Standardisation
MT	Mega Tonne
NAMA	Nationally Appropriate Mitigation Action
NCOS	National Carbon Offset Standard
NGERS	National Greenhouse and Energy Reporting Scheme
NPQ	North Port Quay
PDS	Public Disclosure Summary
RECs	Renewable Energy Certificates
RET	Renewable Energy Target
UNEP	United Nations Environment Program
UNFCCC	United Nations Framework Convention on Climate Change
V2G	Vehicle to Grid
VKT	Vehicle Kilometere Travelled
WBCSD	World Business Council for Sustainable Development
WRI	World Resources Institute
WSUD	Water Sensitive Urban Design

Chapter 1
Addressing Three Wicked Problems

1.1 Introduction

Climate change and the depletion of natural resources, together with a rapidly rising urban population, are three of the great global challenges facing humanity in the twenty-first century. They are each innately interconnected and together remain the most politically complex, capricious and tenacious issues to address, particularly at the international level. The solution, however, is increasingly identified as being best addressed at the local level, specifically within cities and local communities, where action and innovation can occur at a much more rapid rate (Brugmann 2010; Bulkeley et al. 2011; Glaeser 2011; ICLEI 2010; Roseland 2005; UN Habitat 2011).

Cities everywhere are growing, while rural populations are in decline. It is an enormous challenge to house, transport and provide food and services to the three billion urban dwellers within these growing cities. It is also increasing the already significant impact cities are having on our global environment, that is, being a major cause of climate change and resource depletion. Nevertheless, cities are also fundamental in the solution. As cities are likely to be particularly vulnerable to the impacts of both issues, significant motivation exists for them to take action (World Bank 2010). Cities, faced with the dual challenge of climate change mitigation and adaptation, have therefore long been calling for, and taking action on climate change, thereby leading the way in preparing for a carbon- and resource-constrained future (Brugmann 1996; Bulkeley et al. 2011; Roseland 2005).

Cities have the potential to address these wicked problems as they are the centres of innovation and economic growth, which can be the source of new technologies and new ways of living that can decarbonise our planet. Through the process of urbanisation, cities have been having a significant impact on the rate of population growth (UNEP 1996). Family size tends to reduce when people move to cities as a result of better access to a range of services, including family planning and medical services and greater educational opportunities, as well as the increased living costs, which make larger families more costly (Newman 2011; UNEP 1996).

V. Rauland and P. Newman, *Decarbonising Cities*, Green Energy and Technology, DOI 10.1007/978-3-319-15506-7_1

As more of the world's population move into cities in the coming decades, how cities address their growth, while reducing their climate change impact and resource vulnerabilities, will determine their long-term viability. The way we design, build and adapt our cities, and how we learn to live within them, will play an instrumental role in addressing the key challenges facing us, as cities will ultimately determine our survival as a species on this planet. It is hoped that this book will play a small part in assisting cities in taking up this important role. The following outlines the nature of the three global challenges.

1.2 Climate Change

In 2007, the Intergovernmental Panel on Climate Change[1] released a report that delivered three critical and unequivocal messages, which were the following:

1. The global climate is warming, bringing a range of impacts affecting both the natural and industrial worlds;
2. Humans are very likely a significant cause of this change in climate; and
3. The rate at which the climate is changing, and the subsequent impacts, are occurring much faster than anticipated or projected by the models (IPCC 2007).

It is well understood and accepted that climate change is caused by an increase of carbon dioxide and other greenhouse gases[2] in the earth's atmosphere. While the earth's climate has changed over millions of years due to natural variations in carbon dioxide levels (e.g. through volcanoes and melting permafrost), there is strong evidence correlating the current rise in carbon dioxide with human activities, primarily from the combustion of fossil fuel for energy production and use (IPCC 2007).

Some of the known and predicted impacts of climate change include an increase in floods, droughts and severe storms, sea level rise, species extinction and a spread of diseases (IPCC 2007). However, it is the unknown consequences of climate change, which occur once tipping points are passed leading to a series of cascading and potentially irreversible events, that worry the world's most accomplished scientists and climate experts.

The forecast impacts of climate change are likely to cause significant upheaval in our cities and to our urban infrastructure, disturb our agricultural processes and wreak havoc on natural ecosystems and biodiversity, all at great financial cost to the

[1] This is an international scientific community consisting of thousands of reputable climate scientists and experts from all corners of the globe, coordinated through the United Nations.

[2] Within this book, we will be using the terms carbon, CO_2 and greenhouse gas emissions interchangeably to represent the six GHGs covered under the Kyoto Protocol, unless otherwise stated. The six gases include carbon dioxide (CO_2), methane (CH_4), nitrous oxide (N_2O), hydrofluorocarbons (HFCs), perfluorocarbons (PFCs) and sulphur hexafluoride (SF_6). These gases are represented as tonnes of carbon dioxide equivalent (t CO_2-e).

economy and particularly for future generations (Garnaut 2008; IPCC 2014; Stern 2006; VijayaVenkataRaman et al. 2012). Traditionally, addressing climate change has focused on mitigation, as this was identified as being more cost-effective than paying the largely unknown costs associated with the impacts of climate change (e.g. disaster relief, adaption measures, relocating climate refugees). However, as a certain degree of warming has now been locked into the climate system, greater attention is being given to adaption, particularly in the most vulnerable cities, settlements and areas.[3]

1.3 Resource Depletion and Environmental Degradation

There have been significant concerns for many years regarding the ability of the earth to sustain an ever-increasing population in terms of resources, particularly based on current and predicted extraction rates and consumption trends.

Thomas Malthus was one of the first to warn of the consequences of overconsumption, contending in 1798 that humans will consume resources at an unsustainable rate until an event (i.e. war, famine or disease) will lead to their collapse (Malthus 1999). In 1968, two important publications set the scene for a Malthusian revival. Paul Ehrlich revived the population debate with his book ‘The Population Bomb’ which sold over 2 million copies and Garret Hardin presented the problem of excessive resource consumption as a social dilemma caused by the overuse of common resources in his essay ‘The tragedy of the commons’ (Hardin 1997). However, the resource problem was really bought to the world’s attention in the early 1970s by The Club of Rome through its popular book ‘The Limits to Growth’ (Meadows et al. 1972). The authors explored the relationship between unlimited economic and population growth and a limited supply of resources. They produced several computer-modelled scenarios based on five interconnected variables: world population, food production, industrialisation, resource depletion and pollution. A number of the scenarios tested, including the ‘standard run’, ended in collapse of the earth’s systems (Meadows et al. 1972).

Countless others have also predicted and warned of a variety of global environmental and human catastrophes and collapses resulting from the overconsumption of resources and the inability of the earth’s system to regenerate (Carson 1962; Diamond 2005; Ehrlich et al. 1970; Simon 1981). Although the prophesied ‘extreme’ collapse scenarios have not eventuated, the predictions have not been entirely incorrect. There have been many confirmed peaks and significant declines

[3] Low-lying cities and settlements are at extreme risk, as are cities prone to flooding and drought. Ironically, many of the most vulnerable cities are located in developing countries, which are neither responsible for the problem nor able to financially manage the impacts.

in a variety of natural resources, as well as drops in numbers and extinctions of various species (Meadows et al. 2004; Millennium Ecosystem Assessment (MEA) 2005; Rockström et al. 2009).

Rees and Wackernagle (1996), using their ecological footprint model,[4] have demonstrated that humans are living in a perpetual state of overshoot, meaning more resources are being consumed and more waste being produced than the earth can currently replace or absorb. This poses significant problems into the future.

The arguments, analyses and viewpoints put forward have been helpful in raising awareness and spurring the debate[5] around issues associated with overconsumption and the finite capacity of the earth, particularly in relation to a continuously increasing human population. Few of them, however, suggest much about what cities can do to ease the problems.

1.4 Population Growth

Since world population reached its first billion in the early 1800s, it has been growing at an exponential rate. While it took all of human history till then to reach the first billion, world population has doubled several times since, each in a substantially shorter period of time causing many people to question how long this rapid growth can continue unchecked.

In 2011, our population reached seven billion, reawakening concern over the impacts of population growth and reigniting the debate on what a sustainable population might be. It is worth noting that population in isolation is not the primary problem. It is the impact a population has on its environment, due to a variety of factors such as consumption of resources and production of waste that creates the issue. In an attempt to demonstrate this relational impact, Ehrlich et al. (1970) proposed a formula, which they expressed as IPAT, which means:

$$\text{Impact} = \text{Population} \times \text{Affluence (or consumption)} \times \text{Technology}.$$

The formula shows that a person's impact will vary depending on where they are in the world, how much they can afford and how efficient the infrastructure and technologies servicing them are. While it acknowledges that some people will have a far lower impact on the planet than others, whether from the use of better technology or from having lower affluence, the formula ultimately leads to the assertion that an increase in population unequivocally results in a net negative impact.

[4] An ecological footprint is defined as 'the land and water area that is required to support indefinitely the material standard of living of a given population, using prevailing technology' (Chambers et al. 2000: 17).

[5] Julian Simon contradicted the gloomy perspective held by Ehrlich and others, suggesting that technological advances and human ingenuity would prevail. Simon and Ehrlich had numerous public debates and wagers on this topic.

Others have contested this conclusion, suggesting that humans are also capable of solving many of the world's resource problems, and of improving the overall condition of the planet (Newman 2011; Simon 1981) particularly through better-designed cities (Birkeland 2008; Brugmann 2010; Dodman 2009; Droege 2006; Glaeser 2011; Newman et al. 2009; Newman and Jennings 2008; Simon 1981). Highlighting human ingenuity as the most important natural resource, Simon (1981) in his book 'The Ultimate Resource' contends that humans have a boundless capacity to create and adapt in order to provide and survive, and thus argues that population growth can actually create more resources. It has also been argued that human ingenuity and creative thinking are amplified when people live in close proximity to one another, as ideas and solutions spread and evolve far quicker (Brugmann 2010; Glaeser 2011; Simon 1981). From this premise, one could derive that a growing population living in denser cities could indeed help to provide greater solutions to the global issues faced.

Nevertheless, there are people and international agencies, who advocate for a decreasing population in order to cope with the environmental pressures such as climate change, peak oil and resource scarcities (Ehrlich et al. 1970; O'Connor and Lines 2010; The United Nations Population Fund (UNFPA) 2012). The major challenge, however, is how to achieve such a reduction. Forceful methods to control population growth such as mandatory sterilisation or contraception, and policies that disincentivize multiple children such as China's one child policy, often have atrocious consequences (Ren 1995), and thus remain highly controversial and unlikely to receive universal support. Extremist ways of controlling population throughout history have led to mass genocide and ethnic cleansing by infamous figures such as Hitler and Pol Pot (Newman 2011).

However, there are also less aggressive measures to reduce population that are proving successful and are generally accepted by all sides of the debate. For example, global movements aimed at improving education for women and providing better access to family planning services and contraception, primarily in developing countries where current access to both are limited, have had a valuable and significant impact in reducing population (UNEP 1996).

In terms of developed countries and cities, Newman (2011) argues that limiting population is not only unlikely to do much in addressing the resource and environmental concerns facing cities and the planet, but could in fact produce entirely the opposite effect, which is to ensure that the inefficiencies of the current system are locked in or stabilised in stagnant or declining cities. On the other hand, increasing population on purely economic grounds and using existing and often out-dated and inefficient systems will also result in higher carbon emissions, declining resources and increasing environmental concerns. Newman (2011), therefore, proposes a new approach for addressing the debate, which uses population growth in cities (generally accompanied by economic growth) to enable the implementation of new low-carbon technologies and more intelligent infrastructure to transform cities and the urban environment into healthier, more liveable and productive low-carbon cities. Labelled 'The Urban Sustainability Model', Newman (2011) contends that the three factors highlighted by Ehrlich et al. (1970)—population, affluence and technology—are useful to consider with regard to our impact on the planet, but do not inevitably result in a worsening impact.

While it is not the intention of this book to provide a comprehensive or definitive argument or opinion on the population issue, it is based on the premise that humans can, through well-designed cities that include new technologies and new ways of living, create a more sustainable world and future.

1.5 A Manageable Solution: How Cities Address the Challenges

Despite occupying only around 2.5 % of the planet's landmass,[6] cities consume extraordinary amounts of natural resources and discharge considerable volumes of waste into the atmosphere, waterways, oceans and soil (Barles 2010; Potere and Schneider 2007; UNEP 1996). Numerous studies have shown that the amount of resources required to sustain cities, including the consumption of raw materials and the land required to absorb wastes, often extend far beyond a city's geographic or legislative boundary (Churkina 2008; Hoornweg et al. 2011; Kennedy et al. 2011; Satterthwaite 2008). As a result, cities, through their local activities and global consumption habits, are having an increasingly wider impact on the global environment (Borghesi and Vercelli 2003), most notably in the form of anthropocentric climate change. Some research has suggested that up to 80 percentage of global GHG emissions are attributable to cities (Satterthwaite 2008), though there is still debate on the exact contribution (Dodman 2009; Hoornweg et al. 2011). However, on a per capita basis, cities can also be shown to have a smaller carbon footprint than their rural counterparts (Dodman 2009), depending on how carbon is attributed.

Cities are also likely to be the most severely affected and vulnerable to the effects of climate change[7] and resource depletion. On the other hand, cities have also been identified as having the greatest ability to address environmental problems, through improved infrastructure provisions, local action planning and capacity to generate and accelerate innovation (Brugmann 2010; ICLEI 2010; World Bank 2010). They are also where education and health services can be made available to all. Recognising that overconsumption and inefficiency are largely designed into our cities through our infrastructure and service provision (Bartholomew and Ewing 2009; Ewing et al. 2008; Newman and Kenworthy 1999; Newton 2012), greater attention is now being given to enabling low-carbon transitions in urban planning that foster greater flexibility in systems and urban design (Newton 2008).

---

[6] This percentage is increasing rapidly as more people move into cities as a result of urbanisation and is expected to grow considerably if cities grow through urban sprawl instead of density.

[7] For example, many cities were built on low-lying land at the mouth or rivers, which leave them very susceptible to sea level rise.

Thus, cities, through their ability to redesign themselves, their capacity to innovate, provide employment, health and education services and their unique decision making capabilities, can dramatically affect not only population growth, but also the amount of energy and resources consumed and emissions produced, making them a powerful force in global GHG mitigation (Dhakal 2010; Dhakal and Shrestha 2010; Dodman 2009; Ewing et al. 2008; Hoornweg et al. 2011; Satterthwaite 2008).

1.6 Aim of the Book

The aim of this book is to bring greater understanding around the total carbon reduction potential of urban development and identify what processes and opportunities exist for encouraging wider adoption of low-carbon urban development. The book highlights the key emissions' sources associated with urban development and works them into a framework, which if adopted, could help to bring greater standardisation and comparability between carbon claims. Carbon certification is discussed as a way to acknowledge and reward progressive developers, as well as being a potential mechanism for financing innovative low-carbon urban development.

While shedding light on some of the main barriers associated with low-carbon urban development, the book aims to highlight the abundant opportunities to overcome them, thus promoting wider adoption of the concept. Ultimately, it is hoped that the framework proposed in this book could help to lay the foundation for carbon trading within the built environment, including the potential development of new carbon markets that could transform the urban development process and help to mainstream the concept of low-carbon and carbon-neutral urban development.

Overall, this book aims to present a positive approach on how the decarbonising process can be achieved at a local level on a global scale.

1.7 Structure of Book

Chapter 2: The Global Shift to a Low-Carbon Economy Chapter 2 discusses broadly how we target carbon within economies. It distinguishes between what the authors term the 'Front-End' of the economy and the 'End-User' and highlights the benefits and opportunities available from targeting the End-User through voluntary carbon abatement at the urban level.

Chapter 3: Why Cities? Chapter 3 begins by highlighting cities' vulnerability to climate change and why they are keen to take action. An overview of the research relating to how greenhouse gas emissions are attributed to cities globally is then provided. The potential of cities to address climate change is discussed in relation to a number of factors that can influence the amount of carbon associated with cities.

Australia is used as an example to highlight these main sources of carbon within the built environment, which includes a discussion on urban form and an overview of the current arguments for and against various types of urban form and their ability to reduce emissions.

Chapter 4: Low-Carbon Resource Management in Cities This chapter highlights some of the main issues associated with the current large, centralised systems for managing resources in cities and how they contribute to greenhouse gas emissions. Several key opportunities to reduce emissions within our cities and built environment are identified, many of which are based around decentralised approaches to managing energy, water and waste. These are then discussed in relation to Green Infrastructure and how such systems and approaches can begin to close loops. The approaches discussed in this chapter help to inform the development of the carbon accounting framework discussed in Chaps. 11 and 12.

Chapter 5: The Precinct—The New Scale for Decarbonising Chapter 5 takes a look at the precinct scale, defines what is meant by precinct and examines why this level is identified as the optimal scale for GHG abatement.

Chapter 6: Eco-Precincts This chapter begins with an examination of the concept of eco-cities. It analyses their emergence, growth and their evolution into the new low- and carbon-neutral cities of the twenty-first century. A review of nine eco-cities and low-carbon developments is provided: six international case studies and three Australian case studies. The objective of this chapter is to better understand the areas targeted within each city or development, in terms of emissions, and how they compare with each other. The case studies are analysed in terms of their eco-claim, low-carbon initiatives and overall success in achieving their goal.

Chapter 7: The Rise of Carbon Neutrality Chapter 7 examines the effect the rise in voluntary action is having when addressing climate change, particularly regarding the demand for 'green', low-carbon and carbon-neutral products. Issues around 'greenwashing' and the reliability of carbon claims are discussed. Various definitions of carbon neutrality are provided along with a discussion of the key issues that need to be considered when making carbon claims. Offsets are also examined, as they form an integral part of achieving carbon neutrality, and are the reason for the significant growth in the voluntary carbon market.

Chapter 8: Counting Carbon in Cities This chapter begins with a discussion of key literature pertaining to GHG accounting at the city scale. Several of the more prominent existing and proposed city-scale GHG methodologies, tools and inventories are identified, and a brief description of the most recognised and utilised of these frameworks and initiatives is provided. This is followed by a brief overview of the current academic discourse around how carbon should be accounted for in cities, which focuses on boundaries.

Chapter 9: Rating Carbon in Urban Development Chapter 9 provides a summary of existing tools that target GHG emissions at the precinct scale. Some of the most well-known frameworks and tools are examined in terms of their relevance and suitability for calculating precinct-scale emissions. A range of issues are identified and discussed, highlighting some of the key benefits and limitations of current tools.

Chapter 10: Certifying Carbon-Neutral Urban Development Chapter 10 investigates whether certification can lead to greater credibility in carbon claims, highlighting issues pertaining to the Australian Competition and Consumer Commission (ACCC). An overview of the organisational structures and key documents underpinning the certification industry is then provided, followed by a brief summary of several well-known existing Australian and international carbon certification schemes.

Chapter 11: Spotlight: The Australian Carbon Neutral Standard Australia's National Carbon Offset Standard Carbon Neutral Program (NCOS-CNP) is identified as one of the leading independent certification schemes worldwide and is thus discussed in greater depth in this chapter. This chapter identifies and discusses several potential issues and challenges for urban development to achieve certification under this programme, thus providing insight and tips for future programmes.

Chapter 12: A New Framework and Core Elements A new framework for calculating the emissions associated with precinct-scale development is proposed in this chapter, based on the carbon abatement opportunities identified in Chaps. 3 and 4. Guidance is provided on how to maintain carbon neutrality, or the desired carbon goal, within the context of urban development, in order to ensure carbon claims remain relevant and accurate. As the framework is largely based around Australia's National Carbon Offset Standard Carbon Neutral Program, this chapter addresses the unique issues and challenges identified in Chap. 11 that may prevent urban development from pursuing certification.

Chapter 13: Making It Work The primary barriers preventing greater uptake of low-carbon land development within cities are analysed in this chapter. Various opportunities are then highlighted which could assist in overcoming some of these barriers, many of which are underpinned by the need for a reliable system for determining eligibility for the incentives. A brief investigation is undertaken on whether certification could be used, not only as a way to acknowledge the carbon reduction potential of urban development, but also as a mechanism to determine a development's eligibility to receive incentives.

Chapter 14: Final Thoughts The final chapter highlights the key learning from this book, particularly around urban development's potential to reduce emissions, and the importance of acknowledging the role of urban development in global decarbonisation. Recommendations on how this research can help to inform policy are presented, particularly around issues of implementation of a certification scheme for urban development. Ideas for future work are identified, which include how cities can be further engaged and encouraged to design and construct low-carbon developments.

References

Barles, S. (2010). Society, energy and materials: The contribution of urban metabolism studies to sustainable urban development issues. *Journal of Environmental Planning and Management, 43*(4), 439–455.

Bartholomew, K., & Ewing, R. (2009). Land use-transportation scenario planning: A meta-analysis. *Journal of the American Planning Association, 75*(1), 1–15.

Birkeland, J. (2008). *Positive development: From vicious circles to virtuous cycles through built environment design*. London: Earthscan.

Borghesi, S., & Vercelli, A. (2003). Sustainable globalisation. *Ecological Economics, 44*, 77–89.

Brugmann, J. (1996). Planning for sustainability at the local government level. *Environmental Impact Assessment Review, 16*(4–6), 363–379.

Brugmann, J. (2010). *Welcome to the urban revolution: How cities are changing the world*. New York: Bloomsbury Press.

Bulkeley, H., Broto, V. C., Hodson, M., & Marvin, S. (2011). *Cities and low carbon transitions*. New York: Routledge.

Carson, R. (1962). *Silent spring*. Boston: Houghton Mifflin.

Chambers, N., Simmons, C., & Wackernagle, M. (2000). *Sharing nature's interest: Ecological footprints as an indicator of sustainability*. London: Earthscan Publications Ltd.

Churkina, G. (2008). Modeling the carbon cycle of urban systems. *Ecological Modelling, 216*, 107–113. Indolence.

Dhakal, S. (2010). GHG emissions from urbanisation and opportunities for urban carbon mitigation. *Current Opinion in Environmental Sustainability, 2*, 277–283.

Dhakal, S., & Shrestha, R. M. (2010). Editorial: Bridging the research gaps for carbon emissions and their management in cities. *Energy Policy, 38*, 4753–4755.

Diamond, J. (2005). *Collapse: How societies choose to fail or succeed*. New York: Viking Books.

Dodman, D. (2009). Blaming cities for climate change? An analysis of urban greenhouse gas emissions inventories. *Environment and Urbanization, 21*, 185–201.

Droege, P. (2006). *The renewable city*. Singapore: Wiley.

Ehrlich, P. R., Erhlich, A. H., & Holdren, J. P. (1970). *Population, resources and environment: Issues in human ecology*. San Francisco: W. H. Freeman & Company.

Ewing, R., Bartholomew, K., Winkelman, S., Walters, J., & Chen, D. (2008). *Growing cooler: The evidence on urban development and climate change*. Washington, DC: Urban Land Institute.

Garnaut, R. (2008). *The garnaut climate change review—final report*. Melbourne: Cambridge University Press.

Glaeser, E. (2011). *The triumph of the city: How our greatest invention makes US richer, smarter, greener, healthier and happier*. Oxford: Macmillan.

Habitat, U. N. (2011). *Cities and climate change: Global report on human settlements 2011*. London: Earthscan.

Hardin, G. (1997). Tragedy of the Commons. In N. Nelissen, J. V. D. Straaten, & L. Klinkers (Eds.), *Classics in environmental studies: An overview of classic texts in environmental studies* (pp. 101–114). Utrecht: International Books.

Hoornweg, D., Sugar, L., & Gomez, C. L. T. (2011). Cities and greenhouse gas emissions: moving forward. *Environment and Urbanization, 23*(1), 207–227.

ICLEI. (2010). Cities in a post-2012 climate policy framework. In *ICLEI Global Reports*. Retrieved from http://ccsl.iccip.net/cities_in_a_post_2012_policy_framework.pdf.

Intergovernmental Panel on Climate Change (IPCC). (2007). Climate change 2007: Mitigation of climate change. In B. Metz, O. Davidson, P. Bosch, R. Dave, & L. Meyer (Eds.), *Contribution of Working Group III to the Fourth Assessment Report of the Intergovernmental Panel on Climate Change*. Cambridge: Cambridge University Press.

Intergovernmental Panel on Climate Change (IPCC). (2014). IPCC Working Group 3: Mitigation of Climate Change, *Assessment Report 5 of Working Group 3*. Geneva: Intergovernmental Panel on Climate Change.

Kennedy, C., Pincetl, S., & Bunje, P. (2011). The study of urban metabolism and its applications to urban planning and design. *Environmental Pollution, 159*, 1965–1973.

Malthus, T. R. (1999). An essay on the principle of population. In G. Gilbert (Ed.), *Oxford World's Classics reprint*. New York: Oxford University Press.

Meadows, D. H., Meadows, D. L., Randers, J., & Behrens, W. W. (1972). *The limits to growth, a global challenge: A report for the club of Rome project on the predicament of mankind*. New York: Universe Books.

Meadows, D. H., Randers, J., & Meadows, D. (2004). *Limits to growth: The 30-Year update*. White River Junction, VT: Chelsea Green Publishing Company.

Millenium Ecosystem Assessment (MEA). (2005). *Ecosystems and human well-being: Synthesis*. Washington, DC: Island Press.

Newman, P. (2011). Sustaining our future: Resolving the conflict over population models. In *19th International Congress on Modeling and Simulation (MODSIM), December 12–16 2011*. Perth, WA: Modelling and Simulation Society of Australia and New Zealand Inc. (MSSANZ).

Newman, P., & Jennings, I. (2008). *Cities as sustainable ecosystems: Principles and practices*. Washington, DC: Island Press.

Newman, P., & Kenworthy, J. (1999). *Sustainability and cities: Overcoming automobile dependence*. Washington, DC: Island Press.

Newman, P., Beatley, T., & Boyer, H. (2009). *Resilient cities: Responding to peak oil and climate change*. Washington, DC: Island Press.

Newton, P. (Ed.). (2008). *Transitions: Pathways towards sustainable urban development in Australia*. Melbourne: CSIRO Publishing.

Newton, P. W. (2012). Liveable and sustainable? Socio-technical challenges for twenty-first-century cities. *Journal of Urban Technology, 19*(1), 81–102.

O'Connor, M., & Lines, W. J. (2010). *Overloading Australia: How governments and media dither and deny on population*. NSW: Envirobook.

Potere, D., & Schneider, A. (2007). A critical look at representations of urban areas in global maps. *GeoJournal, 69*, 55–80.

Ren, X. S. (1995). Sex differences in infant and child mortality in three provinces in China. *Social Science and Medicine, 40*(9), 1259–1269.

Rees, W., & Wackernagle, M. (1996). Urban ecological footprints: Why cities cannot be sustainable—and why they are key to sustainability. *Environmental Impact Assessment Review, 16*, 223–248.

Rockström, J., Steffen, W., Noone, K., Persson, A., Chapin, F. S., Lambin, E., & Foley, J. (2009). Planetary boundaries: Exploring the safe operating space for humanity. *Ecology and Society, 14*(2), 32.

Roseland, M. (2005). *Toward sustainable communities: Resources for citizens and their governments*. Canada: New Society Publishers.

Satterthwaite, D. (2008). Cities contribution to global warming: Notes on the allocation of greenhouse gas emissions. *Environment and Urbanization, 20*, 539–549.

Simon, J. (1981). *The ultimate resource*. Princeton: Princeton University Press.

Stern, N. (2006). *The economics of climate change: The stern review*. Cambridge: Cambridge University Press.

The United Nations Population Fund (UNFPA). (2012). *Population matters for sustainable development*. Retrieved from http://www.unfpa.org/webdav/site/global/shared/documents/publications/2012/UNFPA%20Population%20matters%20for%20sustainable%20development.pdf.

UNEP. (1996). *An urbanizing world: Global report on human settlements 1996*. Oxford: Oxford University Press.

VijayaVenkataRaman, S., Iniyanb, S., & Goic, R. (2012). A review of climate change, mitigation and adaptation. *Renewable and Sustainable Energy Reviews, 16*, 878–897.

World Bank. (2010). *Cities and climate change: An urgent Agenda* (Vol. 10). Washington, DC: Urban Development and Local Government.

Chapter 2
The Global Shift to a Low-Carbon Economy

2.1 Introduction

Scientists have called for 80 % reduction of carbon in the global economy by 2050 in order to keep global warming below two degrees of pre-industrial levels (IPCC 2007, 2014).

Over 90 developed and developing countries, representing around 80 % of global GHG emissions, have now made pledges to reduce their domestic emissions (IPCC 2014).[1] Many of these countries have communicated their existing or proposed emissions reduction actions and measures that will assist them in achieving their targets. Carbon taxes and emissions trading schemes appear to be the most popular among developed countries with legally binding commitments (DIICCSRTE 2013b), while developing countries without binding commitments have, until recently, identified a range of other 'nationally appropriate mitigation actions' (NAMAs), such as energy efficiency, improved forest management and increasing the amount of renewable energy generation (IPCC 2014). However, various cities and provinces in China and India are now considering carbon taxes and trading schemes and have recently exerted greater global leadership by committing to reductions in GHGs that are based on considerable investments in renewables, electric train systems and greater efficiencies in their cities (Newman and Kenworthy 2015).

---

[1] See http://unfccc.int/focus/mitigation/items/7169.php.

V. Rauland and P. Newman, *Decarbonising Cities*, Green Energy and Technology, DOI 10.1007/978-3-319-15506-7_2

2.2 Tackling Emissions: The Front-End Versus End-User

While there are countless ways to tackle carbon emissions, action is commonly split between two broad areas of the economy—the 'Front-End' and the 'End-User' (Newman and Ingvarson 2012). The Front-End generally refers to emissions that are produced from activities that use fossil fuels directly, such as power generation, refining and large industrial practices. Emissions are thus targeted as they enter the economy. Policies focused on the End-User target carbon as it is used in buildings (homes and businesses), as well as emissions from the built environment related to transporting and servicing these land uses.

Table 2.1 below identifies the primary policies and mechanisms for targeting emissions at both ends of the economy, which can be voluntary or mandatory. Examples of these are provided within an Australian context, though are relevant to most cities, particularly in the developed world.[2]

Although the uptake of carbon policies around the world has been a relatively slow process, those that have been implemented have primarily targeted the Front-End of the economy. The reason for this focus is the greater ability to account for and regulate all the carbon entering an economy. This is the approach advocated by most governments, scientists and international NGOs working in the climate space. However, as demand for products and services comes from the End-User, it has become increasingly evident that a policy gap is emerging around the emissions associated with the consumer, which have been growing steadily in line with rising private energy and material consumption. Unsurprisingly, the majority of End-User consumers are located in cities.

Frustration at the inertia and inaction occurring within higher levels of government and at the international level has seen an increasing number of individuals, businesses, local governments and whole cities taking voluntary action to reduce their greenhouse gas emissions at the End-User level, particularly as this also generally leads to improved performance and a reduction in operational costs.

2.3 End-User Carbon Abatement

There has been a significant increase over recent years in individuals, businesses, local governments and nations taking voluntary action to reduce greenhouse gas emissions—mostly at the End-User side of the economy. The motivation for voluntary action may vary from an intrinsic desire to protect the planet and the environment, to promoting corporate social responsibility (CSR) or simply wanting to take a lead in progressing climate change mitigation, often spurred on by the lack

[2] This book contains primarily Australian examples, in order to provide a consistent political and legal framework, particularly in relation to the development of a certification scheme.

Table 2.1 Front-End and End-User carbon mechanisms with Australian examples

Policies and mechanisms	Front-end examples	End-user examples
Regulation	• Mandatory emissions reporting, i.e. National Greenhouse and Energy Reporting Scheme (NGERS) • Renewable energy target (RET) • Clean coal requirements	• Building codes • Mandatory energy disclosure for buildings • Mandatory energy performance standards
Market-based instruments	• Emission trading schemes • Carbon tax • Other tax incentives • Clean development mechanism • Joint implementation	• Voluntary carbon trading • Carbon offset standards and carbon-neutral programs • Subsidies and rebates (e.g. solar panels, home insulation)
Moral pressure	• International treaties • Corporate social responsibility (large businesses and utilities)	• Carbon-neutral businesses, local government areas (LGAs) • Corporate social responsibility (CSR) • Green building council ratings • Other environmental rating schemes • GreenPower^{TM}, NaturalPower^{TM}
Infrastructure and services	• Smart grids • Renewables linked to grids • Public transport • Solar flagship program	• Smart meters • Electric vehicle plug-in facilities • Infrastructure for walking/cycling/transit • Green infrastructure for local developments • Power utility energy efficiency programs
Education	• Energy efficiency programs • Solar cities program	• Household sustainability audits • Travel smart programs • Living smart programs
R&D and demonstration	• Investment in carbon capture and storage (CCS) and clean coal • Solar flagships • Renewable natural gas	• Carbon-neutral communities • Design models for low-carbon planning and building of precincts
Visioning/strategic planning/governance	• Garnaut report and energy white paper • The Australian Federal government's energy efficiency plan	• Strategic planning • Clinton climate initiative • ICLEI cities • Coalition of Australian Governments (COAG) urban planning focus on reducing carbon

Adopted from Newman and Ingvarson (2012)

of action taken by governments worldwide (Rowell 2010). There are also households and businesses that can see distinct economic advantages in reducing their power and fuel consumption (Newton and Newman 2013).

End-User climate change action can take many forms. It can be simple, inexpensive measures such as behaviour change initiatives like turning off lights and cycling to work, or it can be more investment-oriented action such as purchasing GreenPower™ or investing in energy-efficient technologies (Newman et al. 2009).

Another voluntary emission abatement option that has grown significantly in popularity is the concept of carbon offsetting. Offsetting is based on the premise that carbon emissions from one activity can be effectively neutralised by investing in another low-carbon or carbon-free activity or carbon sequestration[3] in a different location, where it may be more cost-effective or practicable than reducing emissions onsite from the original activity (Stern 2006). The concept stems from the recognition that climate change is a global issue and greenhouse gas emissions have the same effect on the atmosphere wherever they are produced or abated (Garnaut 2008). One of the most common reasons for purchasing carbon offsets is to claim carbon neutrality.

Chapter 7 examines the rise in popularity of the concept of carbon neutrality as a way to voluntarily address climate change.

2.4 Using Cities to Lead the Decarbonisation Effort

As highlighted in Chap. 1, cities are making a major contribution to global emissions as well as a number of other global challenges. A question that is therefore explored in this book is: could the creation of zero-carbon or carbon-neutral cities and urban development be a mechanism for achieving the carbon reduction required to reach the global targets for 2050? Reasons for targeting cities and urban development are discussed in Chap. 3.

2.5 Recognising Carbon Reduction in Urban Development

This purpose of this book is to bring greater understanding around the total carbon emissions associated with cities and urban development in order to enable more accurate comparisons to be made, and subsequently, to create a process by which low-carbon or carbon-neutral developments can be recognised, certified and potentially rewarded for their carbon reduction.

[3] Carbon sequestration involves activities that absorb carbon emissions from the atmosphere such as tree planting or more advanced systems such as biochar or sequestering CO_2 in marine sediments.

This book, therefore, seeks to better define what low carbon and carbon neutral could mean in the context of the built environment and has developed new ways of recognising it. Creating a standardised framework and approach to quantifying the carbon emissions arising from precinct-scale development is critical. It will not only help to make assertions more meaningful and comparable, but will also make benchmarks possible. Such a framework will be fundamental in determining the overall contribution urban development can have in addressing climate change and where reductions can be best achieved.

It is expected that the following mechanisms will help to encourage greater uptake of low-carbon and carbo neutral development: the creation of a new mechanism to certify and acknowledge the carbon reductions achieved within urban development; incentives to reward progressive developers; and processes to capture and distribute knowledge gained.

At the global scale, this is the first attempt to fully conceptualise carbon-neutral and various forms of carbon reduction within urban development in all of its characteristics. We have focused on Australia, which helps to provide a well-defined legal and political context for the carbon reduction activities. The Australian Federal Government has also shown leadership in recent years having developed carbon mechanisms at both ends of the economy (e.g. a mandatory carbon price mechanism[4] targeting the Front-End of the economy, and a government-endorsed voluntary carbon neutral certification scheme that recognises carbon abatement from End-Users).

However, when Australia's new Federal Coalition Government was elected in 2013 with Tony Abbott as Prime Minister, they began dismantling many of these carbon reduction initiatives. The world looked on with incredulity. While the Coalition Government's new set of policies—the Direct Action Plan—is generally considered to be a much lesser set of initiatives, it opens the door for an End-User scheme with potential to generate carbon credits from the built environment and low-carbon urban development. Under the previous scheme, this would have been seen as double counting. The possibility of creating a parallel end-user-based carbon certification scheme for the built environment will therefore be pursued in this book. Although this is not a new concept (the city of Tokyo has been trialling an emissions trading scheme for buildings and facilities—the first of its kind—since 2010),[5] the framework proposed in this book encompasses many additional sources of emissions.

[4] This was an emissions trading scheme with a fixed price for 3 years, resembling a carbon tax. It has since been repealed and replaced with a new set of policies.

[5] See http://www.citiscope.org/story/2014/tokyo-carbon-market-office-buildings-all-cap-and-not-much-trade.

2.6 Conclusion

There are many schemes being developed to help create low-carbon futures. Any country has the opportunity to be seen as a leader in climate change mitigation and to compete in the global innovation stakes on decarbonising development, which is rapidly becoming a defining point for post-industrial, post-carbon economies. This is particularly significant for countries like Australia, considering its current urban form, which is based largely around low-density, carbon-intensive urban sprawl and inefficient infrastructure. Combining these two perspectives, i.e. urban development and decarbonisation, through an emissions trading scheme, therefore offers great potential for demonstrating how significant change can happen.

References

Department of Industry, Innovation, Climate Change, Science, Research and Tertiary Education (DIICCSRTE). (2013b). Countries acting now. In *Shaping a global solution.* Retrieved February 04, 2013, from http://www.climatechange.gov.au/government/international/global-action-facts-and-fiction/cc-countries-acting-now.aspx.

Garnaut, R. (2008). *The Garnaut climate change review—final report.* Melbourne: Cambridge University Press.

Intergovernmental Panel on Climate Change [IPCC]. (2007). Climate change 2007: Mitigation of climate change. In B. Metz, O. Davidson, P. Bosch, R. Dave, & L. Meyer (Eds.), *Contribution of working group III to the fourth assessment report of the intergovernmental panel on climate change.* Cambridge: Cambridge University Press.

Intergovernmental Panel on Climate Change [IPCC]. (2014). IPCC working group 3: Mitigation of climate change. In *Assessment report 5 of working group 3.* Geneva: Intergovernmental Panel on Climate Change.

Newman, P., Beatley, T., & Boyer, H. (2009). *Resilient cities: Responding to peak oil and climate change.* Washington DC: Island Press.

Newman, P., & Ingvarson, C. (2012). Changing cities. In *2020 Vision for a sustainable society.* Melbourne: Melbourne University Press.

Newman, P., & Kenworthy, J. (2015). *The end of automobile dependence: How cities are moving beyond car-based planning.* Washington DC: Island Press.

Newton, P., & Newman, P. (2013). In *How will your energy get greener? Depends where you live.* Retrieved from http://theconversation.com/how-will-your-energy-get-greener-depends-where-you-live-17839.

Rowell, A. (2010). *Communities, councils and a low-carbon future: What we can do if governments won't.* Devon: Green Books.

Stern, N. (2006). *The economics of climate change: The stern review.* Cambridge: Cambridge University Press.

Chapter 3
Why Cities?

3.1 Introduction

There are many reasons for focusing on cities when attempting to decarbonise our planet. On the one hand, cities are often blamed as being the primary cause of climate change. On the other hand, they are seen as being the key to the solution. The latter will be the emphasis in this book, as cities continue to grow globally offering the opportunity for large-scale change. Making the most of the opportunity provided by this growth seems to be the most exciting agenda to occupy urban professionals in the next 50 years. To not use the growth of cities and their communities as the basis of global climate policy would be to miss one of the great global opportunities facing civilisation.

The chapter begins by outlining the vulnerability of cities to climate change. It then provides an overview of the current GHG contribution from cities. Acknowledgement is given to variations in estimates in the literature, which differ depending on the approach adopted in measuring emissions. The discussion then focuses on why cities are fundamental in addressing climate change. Various aspects of city design and urban development are examined in terms of their impact on GHG emissions. These include urban form, transport, materials used in construction and the various approaches for managing resources such as energy, water and waste. Density and urban form are also highlighted as key factors in achieving this.

3.2 The Vulnerability of Cities to Climate Change

Cities are at great risk and are particularly vulnerable to the impacts of climate change. An increase in extreme weather events and new weather patterns is predicted, and this will have varying effects on cities, depending on their geographic locations. Some of the effects include an increase in precipitation, cyclones,

V. Rauland and P. Newman, *Decarbonising Cities*, Green Energy and Technology, DOI 10.1007/978-3-319-15506-7_3

flooding, glacial melt and thus sea-level rise, as well as drought and extreme heat events (IPCC 2007). These events will have significant social, economic and health impacts on cities, as well as severely affecting built infrastructure and urban management systems (Garnaut 2008; Stern 2006; UN Habitat 2011). As the majority of our cities are based around large-scale infrastructure and management of resources (i.e. energy, water and waste systems), cities are particularly susceptible and vulnerable to climate change events, as disruptions to these systems affect extremely large areas of cities, and thus people (Greenpeace 2005; CSIRO 2009).

However, it is not only the resource management systems that will be affected by the increased frequency and intensity of storm events. The events are likely to cause significant damage to all sorts of infrastructure in cities, such as bridges, buildings, roads, subways and other transportation systems, in addition to sewer systems, transmission lines and mobile networks.

Heat events, which include extended periods of above average temperatures and which can lead to droughts and bushfires, are also expected to become more frequent in certain geographical areas (IPCC 2007). Cities that not designed to deal with these pressures can expect problems such as buckling of railway tracks, increased heat island effects, additional stress on power grids and more frequent power outages due to greater demand for air conditioning and higher levels of heat-related mortality (Hennessy 2011; UN Habitat 2011).

The 2003 heatwave in Europe, Hurricane Katrina in New Orleans in 2005, the 2010 Pakistani floods, the 2010/11 Brisbane floods and Hurricane Sandy in North America in 2012 have all demonstrated how costly, both financially and in terms of human life, such weather events can be on cities, particularly in relation to large-scale infrastructure. Another significant risk to cities involves sea-level rise. Around 13 % of the world's population currently resides in cities that are considered at risk of rising sea levels (UN Habitat 2011). For some cities, adapting infrastructure to cope with this (e.g. building dykes, levees, sea walls and potentially moving buildings to comply with new planning regulations and building codes) will be costly and challenging. For many others, it may mean relocating. Climate refugees are expected to be one of the largest costs associated with climate change (Stern 2006). Cities will need to be prepared to address this challenging, though less frequently discussed, and potentially costly, issue (Garnaut 2008).

3.3 Cities' Contribution to Carbon Emissions

The broad contribution of urban areas[1] to global GHG emissions was outlined in Chap. 1. However, debate continues on their exact contribution (Dodman 2009; Dhakal 2010; Hoornweg et al. 2011; Satterthwaite 2008; World Bank 2010), with estimates varying between 40 and 80 % based on how emissions are attributed or

[1] The terms 'cities' and 'urban areas' will be used interchangeably within this book.

apportioned to cities, for instance, whether a production- or consumption-based approach[2] is adopted. Other factors identified as contributing to the variation in city-based emissions include accounting methodologies, scopes of emissions, emissions sources and how 'urban' is defined (Dhakal 2010; UN Habitat 2011).

Adopting a consumption-based approach towards calculating city emissions is useful for several reasons. It can improve understanding of the differences in emissions between regional towns, which often have a lot of primary industry and manufacturing, compared to larger cities with more people but fewer energy-intensive industries (thus producing a lower per capita footprint). However, the latter cities often consume considerable amounts of the materials and services provided by the regional towns.

This issue is even more pronounced between countries, as a growing proportion of emission-intensive industries are being outsourced to developing nations to create products and material goods that are then imported and consumed within the developed countries (Satterthwaite 2008). Therefore, allocating emissions based on what cities consume appears to provide a far more realistic and accurate picture. While this approach to carbon accounting is discussed widely by urban researchers and is becoming increasingly accepted as the most useful approach, it has not yet really been translated or taken up by the policy-making community (Dhakal 2010; Dhakal and Strestha 2010). Indeed, most policy that targets carbon focuses on the Front-End of the economy, rather than the End-User or consumer, as discussed in Chap. 1.

The single largest source of urban emissions, around 70 %, comes from the combustion of fossil fuels for energy use in buildings, transport, manufacturing and industry (IEA 2008; UN Habitat 2011). This is largely due to the nineteenth- and twentieth-century fossil fuel-based energy technologies underpinning our cities (e.g. coal-fired power generation and oil for fuel used in transport), which together remain the biggest challenges for reducing emissions in cities.

Despite energy being the dominant source of emissions in most cities worldwide, the exact percentage of energy-related emissions varies between each city based on factors such as affluence, the economic base, climate and geography, population, density and urban form within a city (Dodman 2009; Glaeser 2011; Newman and Kenworthy 1999, 2015; Roberts and Grimes 1997; UN Habitat 2011; Walker and Salt 2006). Density and urban form are particularly important factors in determining emissions. Higher density urban form has been linked to lower GHG emissions, largely due to the reduced need for private transport, as well as greater efficiency in the operational energy of more compact buildings (Bartholomew and Ewing 2009; Burgess 2000; Ewing et al. 2008; Glaeser and Kahn 2010; Newman and Kenworthy 1999; UN Habitat 2011). Thus, many larger cities have been shown to have smaller average per capita carbon footprints than their country's national average (Dodman 2009).

---

[2] Production-based approach captures emissions from the Front-End of the economy (i.e. big polluters and energy intensive industries), while the consumption approach captures emissions as they enter the economy and are consumed by the End-User.

3.4 Why Cities Are Fundamental in Tackling Climate Change

Carbon has played an important role in the formation and reformation of cities for centuries. Cheap fossil fuels have been a driving force in the rapid expansion of cities and in shaping our urban form. It has enabled and facilitated the development of car-dependent suburbs and allowed houses to continually increase in size. The availability of cheap energy based on fossil fuels has meant little regard has been paid to things such as energy efficiency in buildings and other critical urban design features that were fundamental elements of older cities.

As a result, abundant carbon abatement opportunities are now being identified at this city level, particularly within the built environment (McKinsey and Company 2010; World Bank 2010; UN Habitat 2011). Consequently, cities and their built form are now identified as a 'vital part of the global response to climate change' (Broto and Bulkeley 2013, p. 92).

Furthermore, a popular graph developed by McKinsey and Company (2010) (A Global Greenhouse Gas Abatement Cost Curve—see Fig. 3.1) demonstrates how many of the global carbon abatement opportunities can be made on a cost neutral or positive basis, that is, the measures will have an immediate net financial benefit on the economy over their life cycle. Most of these measures are from within the built environment. Other analyses show how longer term economic gains can be made from more fundamental shifts towards redevelopment in a more compact city (Trubka et al. 2010a, b, c).

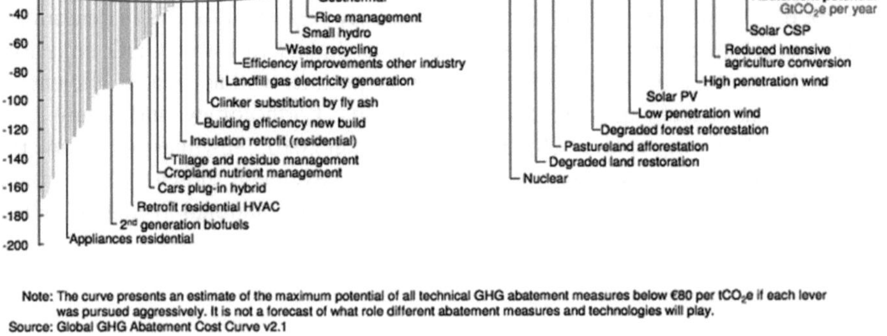

Fig. 3.1 Impact on the financial crisis on carbon economies: version 2.1 of the global greenhouse gas abatement cost curve. *Source* McKinsey and Company (2010) [Reprinted with permission]

Fortunately, cities are well positioned to be able to take action and create change at the local level, particularly in response to climate change and other environmental issues. Being the closest level to the public, cities and local governments have greater capacity to make quick decisions because of the more immediate and effective communication between citizens and local decision-makers, compared to higher levels of government (World Bank 2010). This is critical, as the effects of environmental problems often directly impact cities, demonstrating the need for them to be more pragmatic and swift in their response. As a result, cities have in many cases been leading national governments and international agendas in terms of climate change mitigation and action (Bulkeley et al. 2011; Roseland 2012; UN Habitat 2011).

The dynamic and progressive nature of cities, due to their agglomeration economies, also makes them 'powerhouses' of social, environmental and technological innovation and change (Brugmann 2010; Glaeser 2011; Hollis 2013; Trubka et al. 2010b). Cities are able to foster innovation through proximity and density, which allow the rapid transfer, exchange and development of ideas and knowledge within a small geographical area. Clusters of innovation can learn from each other, and this innovation has resulted in massive leaps in efficiency and reductions in the carbon intensity of urban systems and has provided solutions to many environmental problems over the years (Brugmann 2010; Glaeser 2011; UN Habitat 2011). Acting on climate change at the city level also helps to drive economic competitiveness through increased operational efficiency and investment in new technologies (Brugmann 2010; Sassen 1994).

Proximity and density are also the key to facilitating better public transport infrastructure and decentralised urban resource management options such as co- and tri-generation, as well as providing more efficient and compact housing options (CSIRO 2009; Dodman 2009; Glaeser 2011; Newman and Kenworthy 1999; Rauland and Newman 2011). These aspects of a city (e.g. housing types, transport modes and energy, water and waste infrastructure) are all part of a city's built environment and create its urban form, which ultimately determines its resource consumption patterns and the greenhouse gas emissions associated with it (Ewing et al. 2008; Glaeser 2011; Newton 2012; Roseland 2012). These sources of emissions are discussed later in this chapter.

3.5 Carbon and the Built Environment—An Australian Perspective

Australia currently has one of the largest per capita carbon and ecological footprints in the world (Garnaut 2008; Global Footprint Network 2010). The high carbon emissions are largely a result of the country's overall management of resources, primarily from coal-fired energy generation, as well as the excessive consumption patterns (Australian Conservation Foundation 2007; Newton and Meyer 2012) stemming from the current urban form of Australian cities (Droege 2006; Garnaut 2008; Newman and Kenworthy 2011a; Newton 2008).

3.5.1 Urban Form

Urban form is shaped by transport. Newman and Kenworthy (1999, 2015) show how the Walking City, with its dense, mixed urban form, was based on the ability to walk everywhere in a 1-h travel time budget. This was added to by the Transit City of the nineteenth century where corridors were developed around transit lines based on the same travel time budget. Since the mid-twentieth century, the Automobile City has spread out beyond the 5-km Walking City and the 20-km Transit City to being over 50 km wide. Australian cities have a high proportion of Automobile City areas that are characterised by large, detached, single-family houses situated in low-density, dispersed suburbs (Department of Sustainability, Environment, Water, Population and Communities [DSEWPC] 2010a; Newman and Kenworthy 1999; Newton 2008). Vast amounts of resources, including energy, water and raw materials, are required to sustain these low-density suburban environments. Many studies have demonstrated that higher carbon emissions are associated with low-density, sprawled settlements, compared to higher density, more compact urban form, largely due to electricity consumption and transport emissions (Bartholomew and Ewing 2009; Dodman 2009; Fuller and Crawford 2011; Glaeser 2011; Glaeser and Kahn 2010; Newman and Kenworthy 1999; Newton 1997, 2012; Roseland 2005).

Glaeser and Kahn (2010) found that suburban houses consume up to 88 % more electricity than inner city apartments. This is due to several factors. Firstly, suburban houses, which are predominately stand-alone, detached structures, are less thermally efficient than compact, denser dwellings, which benefit from having shared walls (Ewing et al. 2008; Pitt 2010). Suburban houses are also generally significantly larger than their urban counterparts, thereby requiring significantly more energy for heating and cooling as well as the running of additional appliances (Glaeser and Kahn 2010; Ewing et al. 2008). This is particularly the case for Australia, where new suburban houses have already doubled in size since the 1950s, making them among the largest in the world (Hamilton 2002; Kelly 2010).

Despite improvements in energy efficiency of buildings over the years, research in Australia has shown that per capita energy consumption continues to rise in line with increasing house size (Newton and Meyer 2012). This is being compounded by decreasing occupancy per dwelling, thereby intensifying the need for a greater number of large houses to sustain Australia's growing population (DCCEE 2010; Fuller and Crawford 2011).

Despite cities putting greater focus on infill development, single detached houses built on the fringe still remain the dominant type of new residential development in most of Australia's major cities (Australian Bureau of Statistics (ABS) 2012; DSEWPC 2010b; Newton and Tucker 2011; Newton et al. 2012). Sydney is an exception where infill redevelopment is now much more prevalent than new greenfield development, indicating that the limits to the growth of the Automobile City may have been reached, as in many other large US cities (Newman and Kenworthy 2011b, 2015). Interestingly, studies have shown that greenfield development costs significantly more than infill development when all the additional infrastructure costs are taken into account (Trubka et al. 2010a, b, c; Roseland 2005). However, these costs are not often captured within the price of the land and houses. This type of development is also extremely land intensive and often replaces prime agricultural land or natural, biologically rich areas.

Finally, the embodied energy, or the emissions associated with the materials used in construction, is also greater for larger dwellings. While embodied energy is often excluded from discussions on the emissions contribution of buildings or the built environment, a growing body of research is highlighting the importance of it (see Crawford and Treloar 2005; Sturgis and Roberts 2010). These areas are discussed further below.

3.5.2 Transport

Location and type of housing is major factor contributing to the greenhouse gas emissions associated with our cities (Fuller and Crawford 2011; Glaeser 2011; Glaeser and Kahn 2010; Newman and Kenworthy 1999, 2015; Newman et al. 2009; Naess 1995; Newton 1997). The low-density design of Australian cities, such as many modern cities around the world, has meant they are overwhelmingly car dependent, thus contributing significantly to transport greenhouse gas emissions. Road transport currently accounts for 87 % of Australia's transport emissions with the majority attributed to passenger vehicles (DCCEE 2009).

Extensive research has demonstrated the correlation between density and transport emissions (see Fig. 3.2), showing that as density increases in cities, transport emissions decrease due to a decline in car dependency (Newman and Kenworthy 1999, personal communication with new data; Newman and Kenworthy 2015). This is because public transport uses considerably less energy per person than private vehicles, and public transport is generally only economically viable where sufficient density exists.

While climate change mitigation is an important reason to address Australia's car dependence and ensuing emissions, it is not the only motivation for addressing the issue. Increasing traffic congestion in most major cities, together with rapidly rising fuel prices, are two additional problems that are making the cost of living for people living in sprawled cities increasingly unsustainable.

The economic issues associated with automobile dependence are becoming increasingly clear. An influential report released in 2006 identified the most vulnerable suburbs in Australia's major cities in what was termed the VAMPIRE index, which stood for 'vulnerability assessment for mortgage, petrol and inflation risks and expenditure' (Dodson and Sipe 2006). The study examined the effect that rising mortgages, oil prices and inflation has on people living in different areas within cities. The results showed that lower socio-economic outer suburbs are far more car

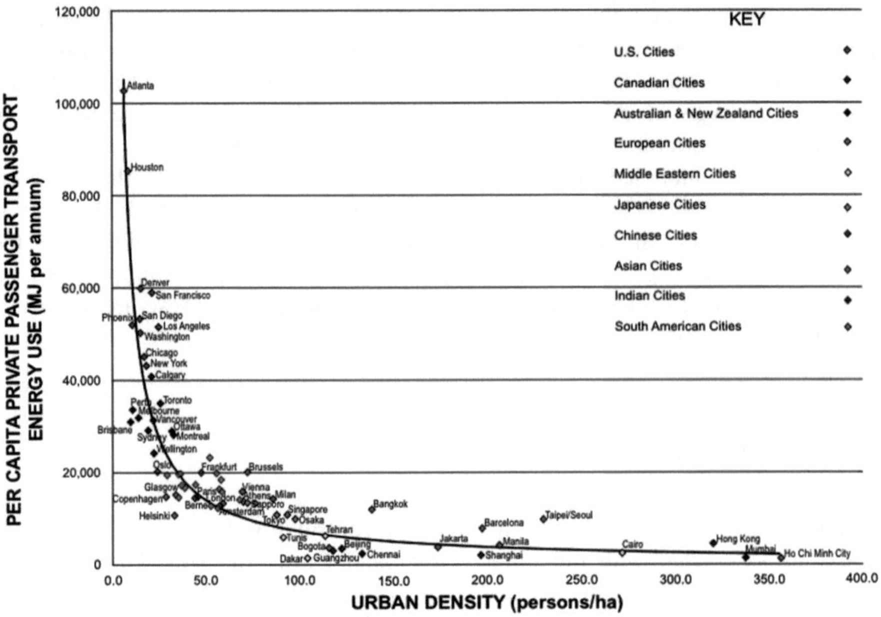

Fig. 3.2 1995 per capita transport energy and urban density. *Source* Kenworthy and Laube (2001)

dependent than inner and middle suburbs and thus are significantly more vulnerable to oil price increases, which is compounded by rising mortgages (Dodson and Sipe 2006). When the GFC occurred, this exact situation happened and many people vulnerable to the rise in fuel prices could not pay their mortgages—a situation that was much more marked in the USA (Newman and Kenworthy 2011b, 2015). Australia's car dependency, particularly in these outer suburbs, is now identified as a key public concern and is becoming an increasingly critical national issue (Department of Infrastructure and Transport 2011).

3.5.2.1 Improving Transport

Ensuring that urban development is designed around, or has sufficient access to, affordable public transport options such as rail (light and heavy) and buses, and that adequate infrastructure is in place to encourage cycling and walking can dramatically reduce the emissions associated with transport in our cities (Newman et al. 2013; Matan and Newman 2012). Encouraging car share schemes to operate within urban areas can also help to reduce car ownership and subsequent vehicle kilometres travelled (VKT) (PMSEIC 2010).[3]

[3] VKT is a common metric for understanding and comparing emissions from transport.

3.5.3 Materials and Construction

The embodied energy[4] (EmE) and subsequent emissions associated with buildings over their lifetime have been estimated to range from as low as 5 % to as high as 62 % of a building's total emissions (Cabeza et al. 2013; Sturgis and Roberts 2010; Thormark 2006). While there are many reasons for this range,[5] research is showing that the proportion of EmE in buildings is increasing as a result of energy efficiency improvements in buildings and the amount of renewable energy used in a building's operation (Cabeza et al. 2013; Dixit et al. 2012). Sturgis and Roberts (2010) note that the proportion of EmE compared to operational energy has been shifting from a general 20:80 ratio to a 40:60 ratio due to these improvements and energy switching and may approach 100 % when new policies, such as the UK's zero emissions housing policy, come into force.

While the percentage may be increasing due to the aforementioned reasons, there are also many opportunities to reduce EmE in materials, which are discussed below.

3.5.3.1 Types of Materials

The EmE of building materials can vary greatly depending on whether materials are new, reused or recycled (see Thormark 2006), where they are sourced, i.e. locally, regionally or internationally (affecting transport emissions), and the type of material used (e.g. wood versus concrete). A study by Preservation Green Laboratory (2011) investigated the differences in the carbon associated with demolition and new construction versus retrofitting/renovating and reusing buildings. It found that 'building reuse almost always yields fewer environmental impacts than new construction when comparing buildings of similar size and functionality' (Preservation Green Lab 2011: vi). This is an important, though perhaps rarely considered, factor when dealing with carbon in the built environment.

Thormark (2006) and Cabeza et al. (2013) both discuss the effect choice of materials has on the embodied carbon associated with buildings, demonstrating that numerous options now exist for materials with a lower EmE component. Buchanan and Levine (1999) argue that low emission construction materials such as wood can

[4] Most studies use the term 'embodied energy' as opposed to 'embodied emissions' as energy has been the typical focus to date. As energy largely correlates with emissions, the term energy is used interchangeably and as a proxy for emissions within this section.

[5] What constitutes 'embodied energy' is currently not well defined or universally agreed upon in existing research. Most studies concur that embodied energy includes the energy associated with the production process of building materials, including the extraction and processing of raw materials, and manufacturing and transport (Cabeza et al. 2013). However, others suggest EmE also includes the onsite construction and demolition processes associated with buildings and final disposal of materials (Dixit et al. 2012). The differences in system boundaries can have a significant effect on the percentage of emissions attributed to EE of buildings.

be substituted for more emission-intensive materials such as aluminium and concrete.

Lehmann (2013) highlights the emissions abatement potential of using low-carbon, prefabricated timber panels to construct residential buildings of up to 10 storeys within an urban infill setting. This method not only cuts GHG emissions considerably through its design and construction, but can also have a significant impact on reducing the costs associated with construction, which can increase the affordability of such apartments (Lehmann 2013). Aye et al. (2012) also identify the emission savings from the use of prefabricated building modules, particularly when they are designed for reuse.

There are also many new ways of reducing the carbon associated with traditionally energy-intensive materials, such as concrete and bricks, by supplementing them with industrial by-products such as fly ash (Cabeza et al. 2013). Furthermore, if energy-intensive materials are made using renewable or carbon-free sources or energy, this could also reduce their embodied energy.[6]

The issue around choice of products or materials with lower EmE is not only relevant to buildings but can also be applied to a range of infrastructure projects within the built environment, including roads, pipes and bridges.

3.5.3.2 The Effect of Improved Operational Performance of Buildings

As the energy performance of buildings improves, due to more efficient design, better insulation and more energy-efficient equipment and appliances, EmE becomes an increasingly important aspect to consider in terms of the building's overall carbon footprint (Cabeza et al. 2013). The more operationally energy efficient a building is, the larger the embodied energy component becomes, and the more important it is to address in order to prevent 'excess negative environmental impacts' (Sturgis and Roberts 2010: 09).

It is worth noting, however, that the increase in operational efficiency can come at a cost to EmE as materials with higher EmE, such as concrete, often provide greater thermal efficiencies (Thormark 2006). It is therefore important to consider both aspects of a building's carbon emissions (i.e. EmE and operational energy) in order to determine the optimal scenario for reducing a building's emissions over its full life cycle.

As EmE has increased in importance, new software and online tools are emerging,[7] which can compare the GHG's associated with different construction materials, providing developers with a quick and robust GHG analysis to help guide decisions.

---

[6] However, Dixit et al. (2012) raise the question of whether this is a reliable way of measuring EE.

[7] See, for example, eTOOL—http://etool.net.au/.

3.5.4 The Role of Energy Efficiency

Twenty-three per cent of Australia's greenhouse gas emissions have been directly attributed to the built environment based on the energy consumed within buildings (Australian Sustainable Built Environment Council [ASBEC] 2008; Garnaut 2008). Fortunately, studies have revealed that this sector also has some of the most cost-effective GHG reductions available within the economy (Climate Works Australia 2010; McKinsey and Company 2008).

A report by the Centre for International Economics (CIE 2008) demonstrated that energy efficiency undertaken within the building sector could reduce GHG emissions by 30–35 % by 2050, even while accommodating new buildings, which they estimate has the potential to reduce the overall cost per tonne of abatement by 14 %.

Some of the vast energy efficiency opportunities within the building sector include improvements to air conditioning in both residential and commercial buildings, residential water heating, insulation, lighting and energy-efficient appliances which have standby features (McKinsey and Company 2008). Retro-fitting these measures will always cost more than installing them up front, so they should always be considered a fundamental element in every new low-carbon development.

3.5.5 Resource Management

Australia's high per capita urban footprints are not only a matter of excessive consumption patterns, but also the way resources are produced and managed. Australian cities, again like many other cities around the world, are predominantly managed by large-scale, centralised infrastructure and systems, many of which use antiquated technologies. These systems are responsible for considerable emissions and consume (or waste) a lot of resources. For example, electricity generation is one of the most water-intensive activities associated with cities, while water is often the largest consumer of energy within cities (particularly if desalination is used to produce potable water). These systems and approaches are discussed further in Chap. 4.

3.5.6 Density and Urban Form

Density refers to the ratio of people or homes to area of land. Well-designed density, which incorporates mixed-use areas (i.e. spaces for living, working and recreational activities), is generally seen to provide numerous sustainability benefits (Jabareen 2006). Density has long been identified as an enabling factor for more efficient public transport infrastructure, as well as increasing the walkability of cities (Newman and Kenworthy 1999, 2015). This can drastically reduce the emissions associated with transport in a city or urban area, as well as improving health by providing opportunities for incidental exercise such as walking and cycling. Encouraging a reduction in car-based transport also reduces the amount of local air pollution, which can reduce the severity of respiratory problems and cardiovascular illness (see Newman and Matan (2012) for an overview of the health implications of transport). Milner et al. (2012) also highlight and discuss many health benefits associated with low-carbon cities.

Increased proximity between home and work and better access to public transport can reduce the time spent commuting, which can also improve lifestyles and productivity at work (Trubka et al. 2010a, b, c). A ground-breaking study on housing in Melbourne revealed that, despite an ongoing preference by Australians for larger houses and gardens, most people were willing to trade off and sacrifice space for greater proximity and accessibility (Kelly 2010).

Increasing building density usually results in a reduction in the size of houses, which can significantly decrease the overall consumption of resources, such as electricity, water and materials (Glaeser 2011; Jabareen 2006). There is some controversy about high-rise causing increases in GHG, but this is not universally accepted (Beattie and Newman 2011). As will be shown in the next chapter, density is also a key to facilitating many of the new low-carbon technologies (e.g. co- and tri-generation) that will be essential in helping to decarbonise our cities (Newman and Kenworthy 2015).

Not all agree with the density attributes discussed above, however. Mees (2010) argues that public transport does not necessarily rely on density, but can be added to suburban neighbourhoods if the right policy mixes are in place to support it. Public attitudes towards density can also pose a problem for developers attempting to build more densely. Negative perceptions of density have a long history, dating back to the 1800s where high density was associated with poor living conditions and poor health (Dempsey et al. 2012). Studies also show some dissatisfaction with the idea of density, suggesting that it is the cause of many neighbourhood problems, such as crime, and that it provides less friendly environments, less access to public open space and less natural lighting, among other factors (Boyko and Cooper 2011). However, many of these issues can and are increasingly being addressed by better design and planning in modern day high-rise developments (Boyko and Cooper 2011; Dempsey et al. 2012).

Sarkissian (cited in Jewell 2013) suggests that engaging the public more in the conversation about density can also help to alleviate worries and improve the chances of creating well-designed, higher density areas being accepted by the community. If the community has been involved in the creation of an urban area, they are less likely to oppose its development.

3.6 Conclusion

The low-density, sprawled urban form based around centralised infrastructure, which has dominated twentieth-century urban planning in the West, remains the focus of countless cities around the world. Australia has been used as an example to illustrate some of the inefficiencies associated with particular car-based city designs. It demonstrated that the current design of Australian cities based on large houses in low-density dispersed car-dependent suburbs is extremely resource and emission intensive.

Many global reports, highlighting the strong interconnection between climate change, urbanisation and infrastructure, have warned of the environmental consequences if new and expanding global cities, particularly in developing countries, adopt traditional twentieth-century low-density city design based around the car and centralised power, water and waste infrastructure (UNEP 1996).

If existing cities around the world continue along a 'business as usual' pathway, and if new cities adopt similar patterns of design, then a significant increase in emissions along with numerous resource concerns can be expected. However, as demonstrated, many opportunities exist to transform cities, which will involve embracing new low-carbon designs and more efficient and resilient urban systems and processes.

References

Australian Bureau of Statistics [ABS]. (2012). *Year Book Australia, 2012*. (Cat. No. 1310.0). Canberra, ACT: ABS. Retrieved from http://www.abs.gov.au/ausstats/abs@.nsf/mf/1301.0.

Australian Conservation Foundation. (2007). *Consuming Australia: Main findings*. Retrieved from http://www.acfonline.org.au/sites/default/files/resource/res_Atlas_Main_Findings.pdf.

Australian Sustainable Built Environment Council [ASBEC]. (2008). *The second plank—Building a low carbon future with energy efficient buildings*. Retrieved from http://www.asbec.asn.au/files/ASBEC%20CCTG%20Second%20Plank%20Report%202.0_0.pdf.

Aye, L., Ngo, T., Crawford, R. H., Gammampila, R., & Mendis, P. (2012). Life cycle greenhouse gas emissions and energy analysis of prefabricated reusable building modules. *Energy and Buildings, 47*, 159–168.

Bartholomew, K., & Ewing, R. (2009). Land use-transportation scenario planning: A meta-analysis. *Journal of the American Planning Association, 75*(1), 1–15.

Beattie, C., & Newman, P. (2011). The density trade-off: Does high rise living contribute more than living in detached dwellings to greenhouse gas emissions? In Whitzman, C & Fincher, R. (Eds.), *State of Australian Cities National Conference,* November 29–December 2, 2011. Melbourne, Vic.: Australian Sustainable Cities and Regions Network.

Boyko, C. T., & Cooper, R. (2011). Clarifying and re-conceptualising density. *Progress in Planning, 76*, 1–61.

Broto, V. C., & Bulkeley, H. (2013). A survey of urban climate change experiments in 100 cities. *Global Environmental Change, 23*, 92–102.

Brugmann, J. (2010). *Welcome to the urban revolution: How cities are changing the world.* New York: Bloomsbury Press.

Buchanan, A., & Levine, S. (1999). Wood based building materials and atmospheric carbon emissions. *Environmental Science & Policy, 2*, 427–437.

Bulkeley, H., Broto, V. C., Hodson, M., & Marvin, S. (2011). *Cities and low carbon transitions*. New York: Routledge.

Burgess, R. (2000). The compact city debate: A global perspective. In M. Jenks & R. Burgess (Eds.), *Compact cities: Sustainable urban forms for developing countries* (pp. 9–24). London: Spon Press.

Cabeza, L. F., Barreneche, C., Miro, L., Morera, J., Bartoli, E., & Fernandez, A. I. (2013). Low carbon and low embodied energy materials in buildings: A review. *Renewable and Sustainable Energy Reviews, 23*, 536–542.

Center for International Economics (CIE). (2008). *The role played by white certificates to combat GHG emissions*. Retrieved from https://www.propertyoz.com.au/wa/library/CIE_White_Certificates_2.pdf.

Climate Works Australia. (2010). *Low carbon growth plan for Australia*. Retrieved from http://www.climateworksaustralia.org/sites/default/files/documents/publications/climateworks_lcgp_australia_full_report_mar2010.pdf.

Crawford, R., & Treloar, G. (2005). An assessment of the energy and water embodied in commercial building construction. In *4th Australian Life Cycle Assessment Conference, Sydney, 2005*. Retrieved from http://www.deakin.edu.au/dro/view/DU:30005640.

CSIRO. (2009). Intelligent grid report: A value proposition for distributed energy in Australia, Commonwealth Scientific and Industrial Research Organisation.

Dempsey, N., Brown, C., & Bramley, G. (2012). The key to sustainable urban development in UK cities? The influence of density on social sustainability. *Progress in Planning, 77*, 89–141.

Department of Climate Change and Energy Efficiency [DCCEE]. (2009). *Australia's national greenhouse accounts: National greenhouse gas inventory accounting for the Kyoto target*. Canberra: Commonwealth of Australia.

Department of Climate Change and Energy Efficiency [DCCEE]. (2010). *Report of the Prime Minister's Task Group on Energy Efficiency*. Retrieved from http://www.industry.gov.au/Energy/EnergyEfficiency/Documents/energy-efficiency/workshops-report.pdf.

Department of Infrastructure and Transport. (2011). *Our cities, our future—A national urban policy for a productive, sustainable and liveable future*. Canberra: Australian Government.

Department of Sustainability, Environment, Water, Population and Communities [DSEWPC]. (2010). *A sustainable population strategy for Australia: Issues paper and appendices*. Canberra: Commonwealth of Australia.

Dhakal, S. (2010). GHG emissions from urbanisation and opportunities for urban carbon mitigation. *Current Opinion in Environmental Sustainability, 2*, 277–283.

Dhakal, S., & Shrestha, R. M. (2010). Editorial: Bridging the research gaps for carbon emissions and their management in cities. *Energy Policy, 38*, 4753–4755.

Dixit, M. K., Fernández-Solís, J. L., Lavy, S., & Culp, C. H. (2012). Need for an embodied energy measurement protocol for buildings: A review paper. *Renewable and Sustainable Energy Reviews, 16*, 3730–3743.

Dodman, D. (2009). Blaming cities for climate change? An analysis of urban greenhouse gas emissions inventories. *Environment and Urbanization, 21*, 185–201.

Dodson, J., & Sipe, N. (2006). *Shocking the suburbs: Urban location, housing debt and oil vulnerability in the Australian city*. Urban Research Program: Research Paper 8. Griffith University. Retrieved from http://www98.griffith.edu.au/dspace/bitstream/handle/10072/12665/41353.pdf;jsessionid=F161B6A7C4D419D970EDD4D18258D905?sequence=1.

Droege, P. (2006). *The renewable city*. Singapore: Wiley.

Ewing, R., Bartholomew, K., Winkelman, S., Walters, J., & Chen, D. (2008). *Growing cooler: The evidence on urban development and climate change*. Washington, DC: Urban Land Institute.

Fuller, R. J., & Crawford, R. H. (2011). Impact of past and future residential housing development patterns on energy demand and related emissions. *Journal of Housing and the Built Environment, 26*, 165–183.

Garnaut, R. (2008). *The garnaut climate change review—Final report*. Melbourne: Cambridge University Press.

Glaeser, E. (2011). *The triumph of the city: How our greatest invention makes US richer, smarter, greener, healthier and happier*. Oxford: Macmillan.

Glaeser, E. L., & Kahn, M. E. (2010). The greenness of cities: Carbon dioxide emissions and urban development. *Journal of Urban Economics, 67*, 404–418.

Global Footprint Network. (2010). *Ecological footprint atlas 2010*. Retrieved from http://www.footprintnetwork.org/images/uploads/Ecological_Footprint_Atlas_2010.pdf.

Greenpeace. (2005). *Decentralising power: An energy revolution for the 21st century*. Retrieved from http://www.greenpeace.org.uk/media/reports/decentralising-power-an-energy-revolution-for-the-21st-century.

Hamilton, C. (2002). *Overconsumption in Australia: The rise of the middle-class battler*. Retrieved from http://www.tai.org.au/documents/dp_fulltext/DP49.pdf.

Hennessy, K. (2011). Climate change impacts. In H. Cleugh, M. S. Smith, M. Battaglia, & P. Graham (Eds.), *Climage change: Science and solutions for Australia*. Collingwood, Victoria: CSIRO.

Hollis, L. (2013). *Cities are good for you: The genius of the metropolis*. New York: Bloomsbury Press.

Hoornweg, D., Sugar, L., & Gomez, C. L. T. (2011). Cities and greenhouse gas emissions: Moving forward. *Environment and Urbanization, 23*(1), 207–227.

IEA. (2008). *World energy outlook*. Paris: International Energy Agency.

Intergovernmental Panel on Climate Change [IPCC]. (2007). Climate change 2007: Mitigation of climate change. In B. Metz, O. Davidson, P. Bosch, R. Dave, & L. Meyer (Eds.), *Contribution of working group III to the 4th assessment report of the intergovernmental panel on climate change*. Cambridge: Cambridge University Press.

Jabareen, Y. R. (2006). Sustainable urban forms: Their typologies, models, and concepts. *Journal of Planning Education and Research, 26*, 38–52.

Jewell, C. (2013). *BPN draws the heavyweights on NSW Planning*. Retrieved from the 5th Estate website: http://www.thefifthestate.com.au/archives/49427/.

Kelly, J. F. (2010). The cities we need. *Grattan institute*. Retrieved from http://grattan.edu.au/wp-content/uploads/2014/04/038_the_cities_we_need.pdf.

Kenworthy, J., & Laube, F. (2001) *The millennium cities database for sustainable transport*. (CDROM Database) International Union (Association) of Public Transport, (UITP), Brussels and Institute for Sustainability and Technology Policy (ISTP), Perth.

Lehmann, S. (2013). Low carbon construction systems using prefabricated engineered solid wood panels for urban infill to significantly reduce greenhouse gas emissions. *Sustainable Cities and Society, 6*, 57–67.

Matan, A., & Newman, P. (2012). Jan Gehl and new visions for walkable Australian cities. *World Transport Policy & Practice, 17*(4), 30–41.

McKinsey & Company. (2008). *An Australian cost curve for greenhouse gas reduction*. Retrieved from http://www.gbca.org.au/docs/McKinsey-Australian_Cost_Curve_for_GHG_Reduction%5B1%5D.pdf.

McKinsey & Company. (2010). *Impact of the financial crisis on carbon economics: Version 2.1 of the global greenhouse gas abatement cost curve*. Retrieved from http://www.mckinsey.com/client_service/sustainability/latest_thinking/greenhouse_gas_abatement_cost_curves.

Mees, P. (2010). *Transport for suburbia: Beyond the automobile age*. London, UK: Earthscan.

Milner, J., Davies, M., & Wilkinson, P. (2012). Urban energy, carbon management (low carbon cities) and co-benefits for human health. *Current Opinion in Environmental Sustainability, 4*, 398–404.

Naess, P. (1995). Central dimensions in a sustainable urban development. *Sustainable Development, 3*, 120–129.

Newman, P., & Kenworthy, J. (1999). *Sustainability and cities: Overcoming automobile dependence*. Washington, DC: Island Press.

Newman, P., & Kenworthy, J. (2011a). Evaluating the transport sector's contribution to greenhouse gas emissions and energy consumption. In R. Salter, S. Dhar & P. Newman (Eds.), *Technologies for climate change mitigation—Transport, TNA Guidebook Series* (pp. 7–23). UNEP Riso Centre for Energy, Climate and Sustainable Development.

Newman, P., & Kenworthy, J. (2011b). Peak car use: Understanding the demise of automobile dependence. *World Transport Policy and Practice, 17*(2), 32–42.

Newman, P., & Kenworthy, J. (2015). *The end of automobile dependence: How cities are moving beyond car-based Planning*. Washington DC: Island Press.

Newman, P., & Matan, A. (2012). Human health and human mobility. *Current Opinion in Environmental Sustainability, 4*(4), 420–426.

Newman, P., Beatley, T., & Boyer, H. (2009). *Resilient cities: Responding to peak oil and climate change*. Washington, DC: Island Press.

Newman, P., Glazebrook, G., & Kenworthy, J. (2013). Peak car and the rise of global rail: Why this is happening and what it means for large and small cities. *Journal of Transportation Technologies, 3*(4), 272–287.

Newton, P. W. (Ed.). (1997). *Re-shaping cities for a more sustainable future—Exploring the link between urban form, air quality, energy & greenhouse gas emissions. Research monograph 6*. Melbourne: AHURI.

Newton, P. (Ed.). (2008). *Transitions: Pathways towards sustainable urban development in Australia*. Melbourne: CSIRO Publishing.

Newton, P. W. (2012). Liveable and sustainable? Socio-technical challenges for twenty-first-century cities. *Journal of Urban Technology, 19*(1), 81–102.

Newton, P., & Meyer, D. (2012). The determinants of urban resource consumption. *Environment and behavior, 44*(1), 107–135.

Newton, P., Newman, P., Glackin, S., & Trubka, S. (2012). Greening the greyfields: Unlocking the development potential of middle suburbs in Australian cities. *World Academy of Science, Engineering and Technology, 71*, 138–157.

Newton, P. W., & Tucker, S. N. (2011). Pathways to decarbonizing the housing sector: A scenario analysis. *Building Research and Information, 39*(1), 34–50.

Pitt, A. (2010). Energy in High Density Cities. In E. Ng (Ed.), *Designing high density cities: For social & environmental sustainability* (pp. 263–271). UK: Earthscan.

PMSEIC. (2010). *Challenges at energy-water-carbon intersections prime minister's science*. Canberra, Australia: Engineering and Innovation Council.

Preservation Green Lab. (2011). *The greenest building: Quantifying the environmental value of building reuse*. Retrieved from: http://www.preservationnation.org/information-center/sustainable-communities/green-lab/lca/The_Greenest_Building_lowres.pdf.

Rauland, V., & Newman, P. (2011). *Decarbonising Australian cities: A new model for creating low carbon, resilient cities*. Paper presented at the MODSIM2011: 19th International Congress on Modelling and Simulation, Perth, WA, December 12–16, 2011.

Roberts, J., & Grimes, P. (1997). Carbon intensity and economic development 1962-1991: A brief exploration if the environmental Kuznets curve. *World Development, 25*(2), 191–198.

Roseland, M. (2005). *Toward sustainable communities: Resources for citizens and their Governments*. Canada: New Society Publishers.

Roseland, M. (2012). *Toward sustainable communities: Solutions for citizens and their governments* (4th ed.). Gabriola Island, BC: New Society Publishers.

Sassen, S. (1994). *Cities and the world economy*. Thousand Oaks, Ca: Pineforge Press.

Satterthwaite, D. (2008). Cities contribution to global warming: notes on the allocation of greenhouse gas emissions. *Environment & Urbanization, 20*, 539–549.

Stern, N. (2006). *The economics of climate change: The stern review*. Cambridge: Cambridge University Press.

Sturgis, S., & Roberts, G. (2010). *Redefining zero: Carbon profiling as a solution to whole life carbon emissions measurement in buildings*. London: RICS Research.

Thormark, C. (2006). The effect of material choice on the total energy need and recycling potential of a building. *Building and Environment, 41*, 1019–1026.

Trubka, R., Newman, P., & Bilsborough, D. (2010a). The costs of urban sprawl—Infrastructure and transportation. *Environment Design Guide, GEN, 83*, 1–6.

Trubka, R., Newman, P., & Bilsborough, D. (2010b). The costs of urban sprawl—Physical activity links to healthcare costs and productivity. *Environment Design Guide, GEN, 85*, 1–13.

Trubka, R., Newman, P., & Bilsborough, D. (2010c). The costs of urban sprawl—Predicting transport greenhouse gases from urban form parameters. *Environment Design Guide, GEN, 84*, 1–16.

UN Habitat. (2011). *Cities and climate change: Global report on human settlements 2011*. London: Earthscan.

UNEP. (1996). *An urbanizing world: Global report on human settlements 1996*. Oxford: Oxford University Press.

Walker, G., & Salt, D. (2006). *Resilience thinking: Sustaining ecosystems and people in a changing world*. Washington, DC: Island Press.

World Bank. (2010). *Cities and climate change: An urgent agenda* (Vol. 10). Washington DC: Urban Development & Local Government.

Chapter 4
Low-Carbon Resource Management in Cities

4.1 Introduction

Cities require access to a variety of natural resources in order to function efficiently and effectively. Energy, water and waste management are critical to ensuring liveability in cities. However, the way these resources are managed can vary substantially between cities and the approach taken can greatly affect the amount of carbon associated with a city.

For many cities around the world, these resources are managed using centralised infrastructure systems. There are often significant emissions associated with these systems. This chapter begins by examining the prevailing technologies and approaches for managing resources, before highlighting alternative, low-carbon options, which can significantly increase the efficiency and thus reduce the amount of carbon associated with cities. These low-carbon options are often developed and managed at the smaller, decentralised scale.

4.2 Centralised Versus Decentralised Management Approaches

It is critical to acknowledge that many of the centralised systems for resource management were developed using nineteenth- and twentieth-century technologies that are not only now outdated and inefficient, but were not originally designed to deal with the problems facing humanity in the twenty-first century, such as climate change and resource depletion.

While great efficiencies were originally associated with 'economies of scale', thus making large-scale systems desirable, many new technologies for distributed generation are now emerging that appear to achieve far greater resource efficiencies

V. Rauland and P. Newman, *Decarbonising Cities*, Green Energy and Technology, DOI 10.1007/978-3-319-15506-7_4

when combined at the local level (Chanan et al. 2009; CSIRO 2009; Newman et al. 2009; Rauland and Newman 2011).

Managing resources more locally can provide numerous benefits. Firstly, generating and managing resources closer to where they are being consumed can radically reduce transport and distribution emissions. Far greater integration between systems can also be achieved at the smaller scale, where for example, outputs from one process (e.g. methane from water treatment) are used as inputs for another (e.g. generating energy). This results in better resource efficiencies, thereby reducing the overall amount of resources required and emissions produced.

In terms of human behaviour and connection to the environment, Giradet (2004) argues that the centralised process for managing resources can leave many people disconnected, unaware of, or having little regard for, the origin or final destination of the natural resources they consume. On the other hand, decentralised processes and systems can bring people much closer to the life systems that support them. The increasing number of individuals, businesses and entire communities embracing opportunities for small-scale production and management of resources such as solar photovoltaic panels, rainwater tanks, community wind farms, biomass co-generation plants and community gardens, are demonstrations of the willingness of people to move beyond the role of passive consumer.

Large-scale centralised processes are extremely vulnerable to disruptions. Bouffard and Kirschen (2008) identify several potential disruptive threats to supply, including terrorist attacks, natural disasters, geopolitical disruptions, ageing of complex infrastructure (e.g. transmission and distribution), climate change and regulatory and economic risk. Decentralised options ensure reliable, localised supply and hence greater security of resources (Ryan 2013). Numerous natural disasters have demonstrated how devastating the widespread failure of power, water and waste systems can be for entire cities. New Zealand provided an unfortunate example where, despite a relatively small area being severely impacted by the 2011 Christchurch earthquake, whole neighbourhoods were left without power, sewerage disposal or clean water. Small scale, decentralised systems, connected to citywide networks, would effectively allow these local areas to continue to function in an island-type capacity. Bouffard and Kirschen (2008) argue that the most efficient network will be flexible and combine the best attributes of both centralised and distributed generation.

Having a larger variety of sources and systems for managing resources also adds diversity to the portfolio of supply options, thereby reducing the risk associated with dependence on a single source. Decentralised systems are also more economically viable and carry less financial risk as they avoid the hefty upfront investment costs associated with large-scale infrastructure such as power stations and dams. As smaller scale options are generally easier and less costly to convert when more efficient alternatives become available, they also help to avoid lock-in (CSIRO 2009).[1]

[1] Lock-in is usually discussed in the context of technology. CSIRO (2009) refers to technology lock-in as a situation when "the dependence of a particular technology [e.g. standardisation of rail gauges] is reinforced by the market it is used in through positive feedback" (p. 219). Lock-in is not always negative, though it can affect or reduce the speed of technological change and innovation.

Cities that are able to substitute outdated, carbon-intensive technologies and processes with new, innovative and more efficient ones, will be helping to solve environmental problems such as climate change and resource depletion. In addition, through these actions, they will be helping to future-proof themselves by increasing their resiliency as outdated technologies become stranded assets as carbon systems rapidly are replaced by renewables (Kennedy and Corfee-Morlort 2013; Newman et al. 2009; Newton 2012; Sinclair 2008).

The following briefly examines some of the most widely adopted approaches for managing of energy, water and waste in cities, and the emissions associated with these sectors, before highlighting some alternative, low-carbon options, which are often based around decentralised technologies. Australia has been used as a case study to demonstrate the inefficiencies and opportunities.

4.2.1 Electricity Generation

Stationary energy production is responsible for around 60 % of Australia's total emissions (DIICCSRTE 2013c).[2] Over 90 % of Australia's electricity is currently supplied through centralised, coal-fired power generation based on a system designed in 1940s (CSIRO 2009; Garnaut 2008; Greenpeace 2005). There were initially several advantages of this energy supply model, including the efficiencies gained from using larger economies of scale, the reduction in transport emissions from having power stations located closer to their fuel source (i.e. coal mines), and the relative ease in transporting the energy to cities through large transmission and distribution lines (CSIRO 2009). However, there are now many issues making this large, centralised model increasingly unsustainable, impractical and a far less appealing option.

It is important to note that coal-fired power generation, which remains the dominant source of energy generation globally, was developed during a time when widespread concern around human induced climate change did not exist. As the science advanced and our understanding grew, it became evident that not only was this power generation extremely damaging to our global environment but it has been, and remains, the primary cause of climate change worldwide and a significant challenge for global climate change mitigation (IPCC 2007).

Coal-fired power generation is particularly carbon intensive for several reasons. To begin with, coal is a particularly dirty fossil fuel when combusted, producing significant emissions. However, the main reason for the significant emissions associated with coal is to do with energy output. Coal-fired power generation is a remarkable inefficient process, whereby around two-thirds of the energy produced during this process (i.e. thermal energy used to spin turbines) is rejected and emitted as steam into the atmosphere. Further losses then occur, as the electricity is

[2] This amount includes fugitive emissions, representing 7.5 % of the total energy sector emissions.

transported vast distances to where it is ultimately consumed within cities. These transmission and distribution (T&D) losses are a result of electricity changing form, usually converted from high to low voltage as it nears the end consumer. Around 7 % of the energy is lost through this process (Dunstan and Langham 2010). Figure 4.1 illustrates the various transformations electricity goes through before it can be used by appliances within households.

As a result of these combined losses, coal-fired power generation is only around 33–35 % efficient (Graus et al. 2007; Greenpeace 2005; Onovwiona and Ugursal 2006). A substantial amount of coal is, therefore, required to generate relatively little energy output. Power stations running on natural gas are slightly more efficient (45 % efficiency) and are less emission intensive as gas is a cleaner burning fuel (Graus et al. 2007). However, the process still discards more than half of the energy produced.

Adding to the resource-intensiveness of this type of generation process is the amount of water required. Electricity generation in Australia is currently responsible for approximately 1.4 % of the nation's water consumption, primarily for producing steam (PMSEIC 2010; Smart and Aspinall 2009). The unreliable nature of Australia's water supply (exacerbated by climate change) is resulting in a leniency towards energy-intensive desalination plants as dependable sources of water. If these trends are any indication of Australia's future water management pathway, the emissions connected to electricity generation (if current technologies prevail) will increase substantially in the future. This relationship between energy and water is discussed further later.

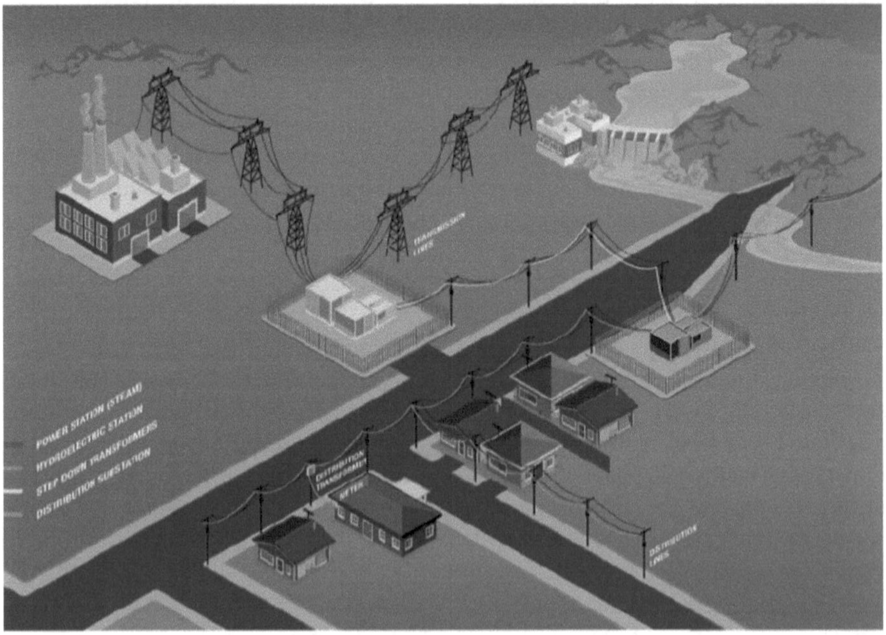

Fig. 4.1 A simplified view of electricity generation and transfer. *Source* CSIRO (2009: 69)

4.2.1.1 Alternative Approaches for Electricity Generation

There are various alternative energy technologies, which can supply cities with less emission-intensive forms of electricity, many of which are from small-scale, distributed energy systems. Solar photovoltaic, solar hot water and small-scale wind are some examples of renewable technologies servicing the individual household or building level, often referred to as micro-generation or distributed generation (Tan et al. 2013).

Options for larger scale and base load carbon-free energy include technologies such as solar thermal, large-scale wind power and geothermal energy. Intermittency in supply is an obvious issue facing these technologies, though many new storage options are becoming available and there is growing interest in this within the multi-residential development arena (Beaudin et al. 2010; Evans et al. 2012; Steffen and Weber 2013). Another increasingly popular option[3] for providing low-carbon, base load energy at a district or precinct scale within cities is co- and tri-generation. While co- and tri-generation can run on bioenergy sources,[4] thereby making them carbon free, currently the technology predominantly uses natural gas, which is still a far cleaner burning fuel than coal, enabling the electricity to be generated within the city without local air pollution.

Co-generation, also known as CHP (combined heat and power), captures the thermal heat energy produced during electricity generation and uses it to provide heating for neighbouring buildings and/or industrial processes. Tri-generation or CCHP (combined cooling, heat and power) adds an absorption chiller to the process to also turn some of the heat energy into coolth, thereby supplying air conditioning as well (City of Sydney 2010; Greenpeace 2005; Jones 2008). Furthermore, as electricity can be generated at, or close to, the point of consumption (i.e. in urban areas), the transmission and distribution losses associated with the generation are drastically reduced.

As a result of these combined efficiencies (i.e. electricity generation, heating and cooling and reduction in transmissions losses), this energy generation process achieves around 75 % efficiency—far greater than large-scale coal-fired generation (City of Sydney 2010; Ge et al. 2009; Greenpeace 2005). Furthermore, as heating and cooling represent between approximately 30 and 40 % of the total energy needs in buildings in Australia (CIE 2008) and just over 50 % globally (WBCSD 2009), significant potential exists for co- and tri-generation to supply much of this demand, which would drastically cut emissions.

Although co- and tri-generation can in theory work at the micro-scale (Pehnt 2008), it is generally more economically viable where there is sufficient building density, and ideally mixed land use, to provide these additional services to (Jaccard

[3] The Borough of Woking, City of London, and City of Sydney have all implemented this technology (or, in Sydney's case, are in the process of installing) as a way to reduce emissions.

[4] Locally produced biomass (i.e. waste from agriculture) and organic municipal waste can be used to create synthetic gas (or syngas), which would provide a carbon free source of electricity.

et al. 1997; Naess 1995). Therefore, in the Australian context, it may mean a shift in urban form from low-density, dispersed neighbourhoods to higher density activity centres, in order to facilitate the application of this transformative technology.

A report released by CSIRO (2009) identified co- and tri-generation, based on natural gas or biomass, as being the most cost-effective, low-carbon distributed energy supply option for Australia.

4.2.1.2 Smart Grids

A smart grid refers to an electricity network that operates much more efficiently than current grids by adopting and integrating information and communication technology (ICT) together with power control devices in appliances that are able to 'talk' to the grid, in order to manage loads more efficiently (PMSEIC 2010). A smart grid provides real-time information on energy costs to encourage End-User efficiency and better user-load management based on price signals.

An important aspect of a smart grid is that it also allows electricity to flow both ways. Enabling the two-way flow of energy will be fundamental to integrating more renewables and decentralised technologies into the grid, and thus being able to better balance intermittent supply with demand (Wissner 2011).

Smart grids will also be essential to enable the storage of electricity. Went et al. (2008) discuss the concept of renewable transport using vehicle-to-grid (V2G) technology, with plug-in hybrid electric vehicles (PHEVs) as a way to store intermittent renewable energy. This form of storage can provide greater opportunities for increased supply of renewables into the system, essentially allowing large-scale renewable energy sources, such as wind power, to provide base load power (Went et al. 2008). EVs combined with smart grids can thus provide a very intelligent solution to current problems with balancing electricity on the grid, particularly around renewable energy integration, as well as potentially providing carbon-free transport. However, in the absence of a smart grid, the increasing adoption of EVs into the market is likely to pose a significant problem as EVs will increase the demand on the current grid system, most likely at peak times (i.e. when people get home from work and start charging EVs).

Smart grids can also enable battery storage in a city. Lithium ion batteries are now becoming dramatically cheaper so households with PV will be able to store energy for periods when the sun is not shining.[5] Battery storage could also be put into neighbourhood substations to provide a stable local supply of energy from larger scale renewables such as wind power in the surrounding region. Such a system of distributed power, with linkages to a larger grid, is likely to emerge as the basis of future urban power. A smart grid will be needed to enable digital control systems to manage the many and various inputs to the electricity grid.

---

[5] See http://www.abc.net.au/news/2014-12-03/demand-for-lithium-batteries-driving-technology-developments/5935472.

4.2.2 *Water Management*

Just as electricity generation is dependent on water, so too is the management of water dependent on energy. As a result, significant greenhouse gas emissions are attributed to the water sector (Rothhausen and Conway 2011). Energy is required throughout the water cycle, from production (i.e. sourcing and treating) through to final disposal. The interdependence between the two systems is often referred to as the 'energy-water nexus' and is becoming an increasingly critical issue due to the 'positive-feedback' loops that are leading to growing vulnerabilities within both systems (Kenway et al. 2008; PMSEIC 2010; Rothhausen and Conway 2011; Siddiqi and Anadon 2011).

The predicted increase in variability of rainfall due to climate change, for instance, is leading to greater demand for more reliable sources of water than current water supplies such as dams and reservoirs. Desalination and wastewater recycling plants are two options that can provide a more dependable supply of water, though both are considerably more energy intensive (Medeazza and Moreau 2007). This means a greater proportion of greenhouse gas emissions are, and will continue to be, attributable to the water supply, which in turn further exacerbates climate change (unless renewable energy is used to power the process). Moreover, as the current production of energy requires significant amounts of water, the increase in energy generation will increase demand for water, which ultimately competes with other urban water needs within our cities (Kenway et al. 2008).

These issues, together with the increasing cost of carbon, are creating concern for the long-term viability of the two systems. Nevertheless, many opportunities and synergies exist that, if integrated in a more holistic way, could increase the resilience of the systems (PMSEIC 2010). Furthermore, if greater focus is given to ensuring that water is only treated to the standard required for use,[6] significant energy-related emissions reductions could be achieved (Chanan et al. 2009).

4.2.2.1 Urban Water System

There are various components of the urban water system, which contribute to greenhouse gas emissions. Griffith-Sattenpiel and Wilson (2009) depict the full water cycle as consisting of five stages, all of which require energy (see Fig. 4.2).

Emissions within the water sector are generated from the production or sourcing of water (i.e. abstraction), the treatment process, distribution (i.e. pumping water to where it is consumed), wastewater treatment and final disposal.

In Australia, this process is primarily managed using a centralised system, whereby water is generally sourced in bulk from a central location, treated to drinking standard, pumped significant distances to where it is consumed,

[6] The percentage of water actually used for potable purposes in urban areas amounts to less than 15 % (Chanan et al. 2009) suggesting that much of this initial treatment process is unnecessary.

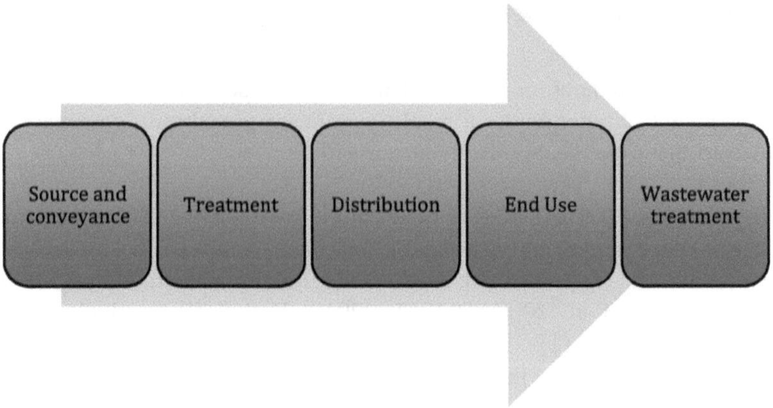

Fig. 4.2 Stages of the water cycle. *Source* adapted from Griffiths-Satternspiel and Wilson (2009)

discharged, and treated again before final disposal, which is typically into rivers and/or out at sea. While the emissions associated with this process are significant, they can vary considerably depending on a range of factors such as topography, location, the water source and treatment method chosen, the level of treatment required (both for water supply and wastewater) and existing infrastructure, i.e. pipe dimensions and design (Kenway et al. 2008). Below is a discussion of the key emission sources associated with various aspects of the water cycle.

4.2.2.2 Dams and Reservoirs

Approximately 80 % of Australia's potable water is currently sourced from dams and reservoirs that use river catchment areas to collect surface water run-off (ABS 2007). This supply option requires the least amount of operational energy (providing it is gravity fed) due to the relatively little treatment required. However, there can be significant embodied emissions associated with the initial construction of reservoirs (i.e. dam walls), as well as the water distribution network, i.e. pipelines (Kenway et al. 2008; Reffold et al. 2008). Furthermore, the initial flooding of large tracts of land in the process of creating a water catchment can produce significant methane emissions as the organic material (i.e. soil and plants) decomposes under water (Hall et al. 2009).

Regardless of the benefits or limitations of reservoir/dam storage or river abstraction, the current and predicted declining rainfall patterns in Australia are making this option less attractive, and in many cases, unviable (ABS 2007).

4.2.2.3 Groundwater

The second largest water supply option (just under one-fifth of Australia's water) comes from groundwater (ABS 2007). This is another comparatively low-energy water supply option, again due to the minimal treatment required to bring water quality to potable standard. However, the emissions associated with this supply option are likely to increase in the future, as increasing energy is required for pumps to extract water from deeper boreholes as water tables decline.[7] Moreover, since groundwater supplies rely on adequate rainfall to replenish aquifers, which again is decreasing due to climate variability, this option is also likely to become less attractive for water utilities and unsustainable in the long term.[8]

4.2.2.4 Desalination

Desalination plants, which convert sea water into potable water for urban use, are becoming an increasingly popular alternative for water utilities that are searching for a culturally acceptable[9] and dependable source of water. Desalination, however, is considerably more energy intensive and therefore emission intensive than traditional water supply options and has some additional environmental impacts associated with this process that are causing some concern (Medeazza and Moreau 2007).

Desalination is at present around four times more energy intensive as conventional sources of water (PMSEIC 2010). If the energy used in the desalination process is sourced from fossil fuels, significant additional emissions can be expected. PMSEIC (2010) note that if all urban water in Australia was sourced from desalination plants, which draw energy from the national electricity grid, national emissions could be expected to rise by around 7.2 $MtCO_2$-e/y or 1.3 % (2010: 41). However, this could be mitigated if desalination plants source their power from renewable energy as has happened in Perth where almost half of the city's water supply now comes from desalination. This is a good example of a city not only adapting to climate change (rainfall reductions have been ascribed to climate change for over a decade) but contributing to mitigation through its urban water policy (Newman 2015; Stocker et al. 2012).

---

[7] Assuming energy is produced using fossil fuels.

[8] There are several trials being run in Australia looking at Managed Aquifer Recharge (MAR) to replenish groundwater supplies. See Bekele et al. (2013).

[9] Compared to reusing wastewater, Fullerton (2001) highlights the difficulty in attaining public acceptance of reusing wastewater.

4.2.2.5 Wastewater Treatment

Significant emissions are associated with the treatment of wastewater due to the energy required during the treatment process and the methane produced from the decomposition of nutrient-rich organic matter. Research conducted by Kenway et al. (2008) demonstrated that each time an additional wastewater treatment process was undertaken (i.e. from primary to secondary and secondary to tertiary), the energy requirement doubled, along with the associated emissions. This could be seen as a deterrent for utilising wastewater as a source of water supply in urban areas due to the high energy demand required for treatment. However, for cities that already require tertiary treatment of water before disposal (e.g. Melbourne), the additional treatment required to bring the water to potable standard is relatively minor compared with the treatment process already undertaken (Kenway et al. 2008).

By-products from treated effluent, which include biogas and biomass, could also be combusted to produce energy, which could be used onsite in the treatment process, thereby reducing demand for emission-intensive grid electricity. In fact, Ainger et al. (2009) suggest that the water sector could become carbon neutral in the future 'by a careful balancing of generation from waste (methane in particular), generation from renewables and the management of its demand patterns' (12).

The high nutrient load in domestic wastewater offers great opportunities for nutrient recovery. By-products such as phosphate, nitrogen and ammonia can be extracted and used as natural fertilisers for agriculture, thereby helping to close resource loops (Giradet 2004; Kenway et al. 2008; Newman and Jennings 2008; van Lier and Lettiga 1999). Installing third pipe systems in urban developments will enable the separation of black water from grey water, which will reduce the energy required in both treatment processes (Kenway et al. 2008).

4.2.2.6 Distribution

The other key contributor to greenhouse gas emissions on the supply side is the pumping and distribution of water through the centralised network within our cities. In some cases, water pumping and distribution can be responsible for the majority of a water utility's greenhouse gas emissions (see Fig. 4.3) for example, where water needs to flow against gravity (Hall et al. 2009; Kenway et al. 2008).

4.2.2.7 End-Use Water Emissions

End-use emissions are generated primarily from the energy used for heating and cooling water in homes and commercial buildings (e.g. for showers, dishwashers, washing machines and kettles). Water efficiency measures and appliances that can reduce the amount of water consumed will directly reduce the amount of energy required to heat the water. Efficient hot water supply systems such as solar or gas hot water systems can also dramatically reduce emissions associated with End-use

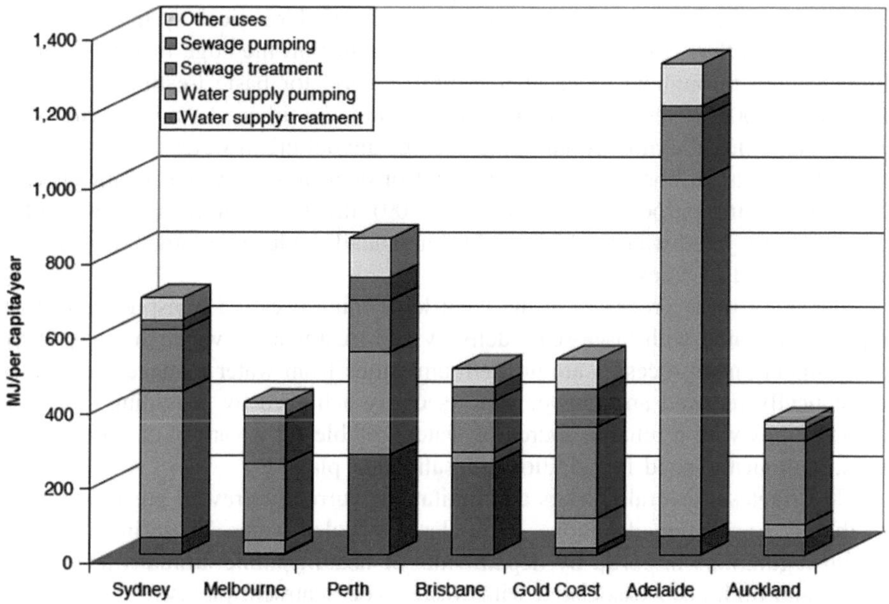

Fig. 4.3 Energy use for water and waste services 2006/07. *Source* Kenway et al. (2008)

water. It is worth noting, however, that the emissions associated with energy in buildings, despite it being connected to water use, are captured within the energy sector's GHG accounting.

Improved demand-side management (e.g. rainwater harvesting at the building scale) can also help to reduce consumption and therefore reduce emissions associated with supply.

4.2.2.8 Alternative Water Management and Urban Design

Research suggests there is significant potential for improving water management within cities by adopting a more decentralised approach that can promote greater integration of processes, help to close loops and increase the sustainability of the system (Newman and Jennings 2008). Chanan et al. (2009) refer to this decentralised approach as a 'soft path' for water management, one that promotes greater integration of demand-side (i.e. End-User efficiency) and supply-side initiatives. The supply side involves sourcing, treating and managing water at a much more local level. This includes dramatically increasing rainwater harvesting and storm water retention within cities, i.e. capturing storm water run-off from rooftops and other impervious surfaces and treating it to the appropriate level to which it will be used, rather than allowing it to flow into the sewer system where it must then undergo an energy-intensive treatment process.

Water-sensitive urban design using natural and biological systems, such as swales and buffers, to collect and filter storm water on the edge of roads, can be implemented throughout cities as a natural method for basic treatment of water. Hatt et al. (2005) notes 'The average annual volume of urban storm water run-off in Australian cities is almost equal to the average annual urban water usage, of which at least 50 % is for non-potable use' (103). Considering most water in cities is used for non-drinking purposes (Chanan et al. 2009), there are significant opportunities to use this recycled water, which could dramatically reduce the amount of scheme water required in cities.

Managing these processes at the local level minimises the transport and distribution associated with large-scale delivery and treatment of water in cities. If the above-mentioned processes are utilised, emissions from water management can be dramatically reduced and greater water security achieved by providing cities and communities with a reliable source of water suitable for a variety of different end uses, without the need for additional desalination plants.

Nevertheless, several barriers and limitations currently prevent greater adoption of the above-mentioned options, particularly recycled water. These include: strict health regulations imposed by departments of health; public attitudes to recycled water; and high costs associated with wastewater treatment plants.

4.2.3 Waste Management

The amount of solid waste produced in Australia is currently around 44 MT per annum, which represents around 11 MT GHG emissions (DEHWA 2010a). The volume of solid waste has doubled over the past 20 years and is growing by approximately 7 % per year (DEHWA 2010b).

4.2.3.1 Landfill

Around half of the materials and products that enter Australian cities exit as waste and are buried in landfill (DEWHA 2010b). This is one of the oldest and most inefficient forms of waste management. It is also particularly emission intensive, as methane (CH_4) is produced by the anaerobic decomposition of organic materials, which occurs for decades after initial waste disposal (Lou and Nair 2009). Methane has a considerably larger impact on climate change than carbon dioxide. There are also many issues surrounding toxic leaching (Ikehata and Liu 2011), particularly into groundwater.

Further emissions are associated with the collection and transportation of waste from cities to landfill sites, which are generally situated outside of cities. This process, therefore, also requires substantial tracts of land, which can remain contaminated and unusable for years after (Ikehata and Liu 2011).

4.2.3.2 Alternative Approaches to Waste Management

Other forms of waste management that produce fewer emissions include waste to energy through incineration and various forms of gasification (Mountouris et al. 2006). However, there are countless options for managing solid waste, which produce varying amounts of GHG emissions depending on the technique adopted. The emissions associated with different treatment strategies have been reviewed in academic literature (Chen and Lin 2008; Dulal and Akbar 2013; Hughes et al. 2006).

Viewing waste as a resource (Tillman 2004) and reusing, recycling and composting as much as possible is the optimal way of dealing with waste (Seadon 2010). This not only has the potential to reduce emissions, but also helps to create a more circular metabolism within our cities as espoused by Wolman (1965) and Giradet (2004). The concept of resource recovery within waste management will no doubt play an essential role in decarbonising cities into the future.

Materials that cannot be reused, recycled or broken down naturally can be treated by new technologies such as plasma arc gasification. This process reduces waste to 1/300 of its original size as well as producing syngas, a renewable form of natural gas (Lombardi et al. 2011). This gas can then be fed directly into a tri-generation system to create low-carbon, renewable energy, thus providing a perfect example of a closed-loop system (City of Sydney 2010). However, such technologies are still relatively rare and therefore expensive.

New waste collection techniques including municipal vacuum waste, which sucks waste through an underground network of pipes to one or several collection points within a city, can also dramatically reduce the transport emissions associated with waste collection, improve the efficiency of recycling and add amenity by reducing the number of garbage trucks on the road (City of Sydney 2010).

The first step in any waste management process, however, should be to minimise or avoid the amount of waste generated in the first instance. In terms of urban development and the construction industry, many opportunities exist to reduce the amount of construction and demolition (C&D) waste being generated. The use of prefabrication for building construction, for example, can dramatically reduce C&D waste entering the waste stream. Approximately 35 % of all the waste that ends up in landfill across Australia comes from C&D waste (DEWHA 2010b). Encouraging society to minimise waste is likely to be more challenging but can be facilitated in urban planning, and design by ensuring sufficient infrastructure is in place to enable good behaviour to occur (e.g. clearly marked recycling points). Behaviour change programs at the local level are also likely to be required.

The very concept of waste is the problem lying at the heart of most cities' inability to adopt more sustainable resource management practices. Waste implies that a material has come to the end of its productive life and no longer has any functional purpose (Tillman 2004). Thermodynamics, especially the concept of entropy, explains how materials become less and less useful when they are passed through a system unless further energy is applied to them. The concept of waste, which does not exist in the natural world, has created the unsustainable linear flow of resources through our cities. However, materials often still contain significant uses that require minimal energy to recycle, though a point is reached where more energy is needed to recycle them than can be saved from the embodied energy in the materials.

With Australia's expected population growth (Productivity Commission 2010) combined with increasing per capita waste production, improving the management of waste within our cities (and most other global cities) is essential and needs to be a national priority.

4.3 Closing Resource Loops

The previous section reveals two important characteristics of centralised systems for resource management. Firstly, they allow resources to flow through cities in a horizontal or linear fashion where consumed resources are extracted as waste (Giradet 2004; Newman and Jennings 2008). And secondly, they often tend to function in isolation, ignoring important synergies and the potential efficiencies achievable when systems are combined. This is vastly different to how natural ecosystems function, in that such systems are based on circular flows and closed-loop systems, where processes are inherently interconnected and resources and nutrients are continuously recycled.

Newman and Kennworthy (1999) suggest that sustainability in cities can be understood in terms of an Extended Metabolism Model where the resource inputs and waste outputs of a city are reduced, while liveability (economic, social and environmental) is increased.

Giradet (2004) argues for a greater system-based approach to urban management that can help to create a more circular metabolism within our cities. Newman et al. (2009) define this metabolic city concept as the 'Eco-Efficient City', which demonstrates how various wastes can be turned into new resources for other processes to radically improve the overall efficiency of systems. Kennedy and Corfee-Morlot (2013) stress the importance of designing low-carbon strategies for city-based infrastructure that takes into account the interdependencies between infrastructure systems.

Many of these processes and systems are increasingly seen to be most effective at the smaller scale, usually at the district or community level (Newman et al. 2009; Newman and Jennings 2008; Roseland 2012). These small-scale, integrated systems form part of a concept defined here as decentralised green infrastructure.

4.4 Green Infrastructure

Green infrastructure is commonly referred to in American urban planning literature as natural or biological infrastructure within cities such as trees, waterways, parks and 'green spaces' (Benedict and McMahon n.d.; Vandermuelen et al. 2011).

Outside the USA, however, the term green infrastructure is applied more broadly and defined as alternative infrastructure options for supplying power and water and treating wastewater and solid waste that can help to achieve sustainability outcomes and reduce a city's carbon footprint (Bunning et al. 2013; Walsh 2011). It can also apply to other aspects of city design such as low-carbon transport infrastructure (i.e. light rail, trains and cycle paths).

In relation to urban resource management services, green infrastructure is typically based on the decentralised or distributed systems mentioned above, which help to create more integrated and resilient infrastructure systems within our cities. Figure 4.4 provides a conceptual model of green infrastructure, demonstrating how interconnected the systems can and should be. It was based on a case study that was conducted as part of the ARC Decarbonising Cities and Regions research project, which underpins much of the research discussed in this book.

The new green infrastructure outlined above is appealing to urban policy makers across the globe wanting to decarbonise their cities. This is especially so in mega cities such as Jakarta and Dhaka which have no centralised water and waste systems and where energy supply is very unreliable; it is possible to leapfrog into the twenty-first-century decentralised green infrastructure through a series of smaller systems, which can then be integrated. There are several factors, however, that must be addressed as part of the change including governance and urban form. New governance models will need to be developed to facilitate decentralised infrastructure

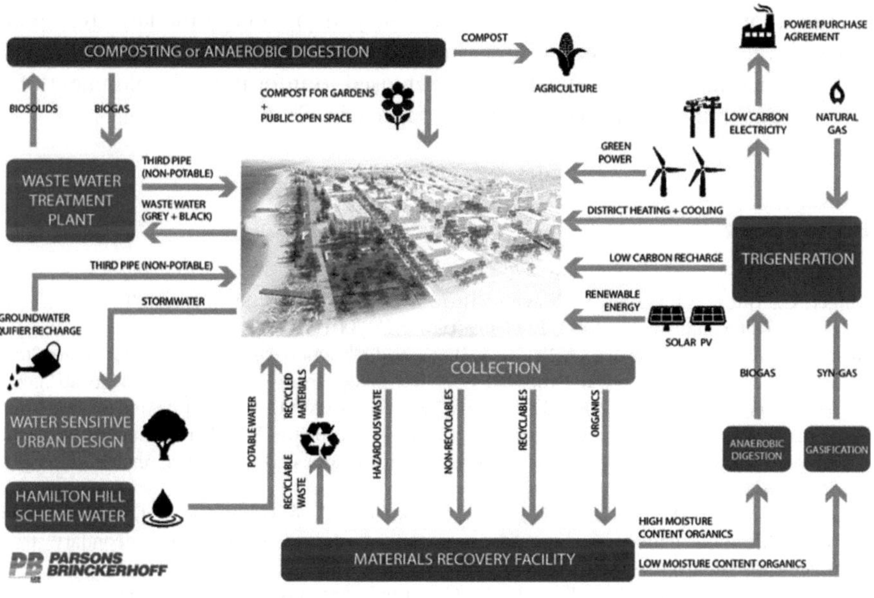

Fig. 4.4 Cockburn Coast Green Infrastructure Eco-Cycle. *Source* Walsh (2011)

(see Bunning et al. 2013). Urban form can also not be separated from these emerging technologies as precinct-scale technologies and systems generally require adequate density and mix to be most efficient and effective, as well as economically viable. This is particularly the case with decentralised energy infrastructure (Naess 1995).

4.5 Conclusion

This chapter analysed the carbon implications associated with various types of resource management infrastructure and systems within cities. The examples highlighted that large-scale, centralised management of resources in cities, which are often based on outdated technologies and compounded by ageing infrastructure, is the cause of significant emissions within cities and indeed, globally.

However, while significant emissions may currently be attributed to cities and the built environment, substantial emissions reductions are also possible if alternative designs and approaches are adopted. It has not been the intention of this chapter to argue for or recommend specific technologies or the most optimal set of solutions for decarbonising cities. Rather, to demonstrate and highlight the significant potential for reducing emissions.

Furthermore, it appears that the density and urban form of a city, as highlighted in Chap. 3, is inherently connected to the green infrastructure described above and will be a crucial part of any decarbonising process.

This chapter, together with Chap. 3, has helped to identify the key areas within urban development and the built environment that offer the greatest potential for carbon reduction. This knowledge has been used to inform the development of the carbon accounting framework discussed further in Chap. 12.

References

Ainger, C., Butler, D. Caffor, I. Crawford-Brown, D. Helm, D. & Stephenson, T. (2009). A low carbon water industry in 2050. *Evidence* (vol. SC070010/R3). Bristol.

Australian Bureau of Statistics [ABS]. (2007). *Australia's environment: issues and trends 2007—Special issue: water.* (Cat No. 4613.0). Retrieved from http://www.ausstats.abs.gov.au/Ausstats/subscriber.nsf/0/8062FE062727E272CA2573AA000F535D/$File/46130_2007.pdf.

Beaudin, M., Zareipour, H., Schellenberglabe, A., & Rosehart, W. (2010). Energy storage for mitigating the variability of renewable electricity sources: An updated review. *Energy for Sustainable Development, 14*, 302–314.

Bekele, E., Toze, S., Patterson, B., Fegg, W., Shackleton, M., & Higginson, S. (2013). Evaluating two infiltration gallery designs for managed aquifer recharge using secondary treated wastewater. *Journal of Environmental Management, 117*, 115–120.

Benedict, M. A., & McMahon, E. T. (n.d.). Green infrastructure: smart conservation for the 21st century. *Sprawl watch clearinghouse monograph series.* Retrieved from http://www.sprawlwatch.org/greeninfrastructure.pdf.

Bouffard, F., & Kirschen, D. S. (2008). Centralised and distributed electricity systems. *Energy Policy, 36*, 4504–4508.
Bunning, J., Beattie, C., Rauland, V., & Newman, P. (2013). Low-carbon sustainable precincts: An Australian perspective. *Sustainability, 5*(6), 2305–2326.
Center for International Economics [CIE]. (2008). The role played by white certificates to combat GHG emissions. Retrieved from https://www.propertyoz.com.au/wa/library/CIE_White_Certificates_2.pdf.
Chanan, A., Kanasamy, S., Vigneswaran, S., & Sharma, D. (2009). A gradualist approach to address Australia's urban water challenge. *Desalination, 249*(3), 1012–1016.
Chen, T. C., & Lin, C. F. (2008). Greenhouse gases emissions from waste management practices using Life Cycle Inventory model. *Journal of Hazardous Materials, 155*, 23–31.
City of Sydney. (2010). *Green infrastructure for Sydney*. Retrieved form http://www.sydney2030.com.au/development-in-2030/city-wide-projects/powering-sydney-allan-jones.
CSIRO. (2009). *Intelligent grid report: A value proposition for distributed energy in Australia*. Commonwealth Scientific and Industrial Research Organisation.
Department of Industry, Innovation, Climate Change, Science, Research and Tertiary Education [DIICCSRTE]. (2013c). *Australian national greenhouse accounts: National inventory report 2011*. Retrieved from http://www.climatechange.gov.au/sites/climatechange/files/documents/05_2013/AUS_NIR_2011_Vol1.pdf.
Department of the Environment, Water, Heritage and the Arts [DEHWA]. (2010a). *National waste policy: Fact sheet*. Commonwealth of Australia. Retrieved from http://www.environment.gov.au/system/files/pages/94aa70c5–6681-44c6-8d83-77606d1d6afe/files/fs-national-waste-policy.pdf.
Department of the Environment, Water, Heritage and the Arts [DEWHA]. (2010b). *National waste report 2010*. Canberra: Commonwealth of Australia. Retrieved from http://www.scew.gov.au/system/files/resources/020c2577-eac9-0494-493c-d1ce2b4442e5/files/wastemgt-nat-waste-report-final-20-fullreport-201005-0.pdf.
Dulal, H. B., & Akbar, S. (2013). Greenhouse gas emission reduction options for cities: Finding the coincidence of agendas between local priorities and climate change mitigation objectives. *Habitat International, 38*, 100–105.
Dunstan, C., & Langham, E. (2010). *Close to home: Potential benefits of decentralised energy for NSW electricity consumers*. Sydney: Institute for Sustainable Futures.
Evans, A., Strezov, V., & Evans, T. J. (2012). Assessment of utility energy storage options for increased renewable energy penetration. *Renewable and Sustainable Energy Reviews, 16*, 4141–4147.
Fullerton, T. (2001). *Watershed: Deciding our water future*. Sydney, NSW: ABC Books.
Garnaut, R. (2008). *The garnaut climate change review—Final report*. Melbourne: Cambridge University Press.
Ge, Y. T., Tassou, S. A., Chaer, I., & Suguartha, N. (2009). Performance evaluation of a tri-generation system with simulation and experiment. *Applied Energy, 86*, 2317–2336.
Giradet, H. (2004). Metabolism of cities: From creating sustainable cities. In Wheeler, S. M & T. Beatley (Eds.), *The sustainable urban development reader* (pp. 125–132). New York: Routledge.
Graus, W. H. J., Voogt, M., & Worrell, E. (2007). International comparison of energy efficiency of fossil power generation. *Energy Policy, 35*, 3936–3951.
Greenpeace. (2005). *Decentralising power: An energy revolution for the 21st century*. Retrieved from http://www.greenpeace.org.uk/media/reports/decentralising-power-an-energy-revolution-for-the-21st-century.
Griffith-Sattenspiel, B., & Wilson, W. (2009). *The carbon footprint of water*. Retrieved from http://www.csu.edu/cerc/researchreports/documents/CarbonFootprintofWater-RiverNetwork-2009.pdf.
Hall, M., West, J. Lane, J. Haas, D. D. & Sherman, B. (2009). *Energy and greenhouse gas emissions for the SEQ water strategy*. Retrieved from http://www.urbanwateralliance.org.au/publications/UWSRA-tr14.pdf.

Hatt, B. E., Deletic, A., & Fletcher, T. D. (2005). Integrated treatment and recycling of stormwater: A review of Australian practice. *Journal of Environmental Management, 79*, 102–113.

Hughes, R., Anda, M., Ho, G., & Mathew, K. (2006). Integrated waste management. In D. Marinova, D. Annandale, & J. Phillimore (Eds.), *International handbook on environmental technology management* (pp. 405–421). Cheltenham: Egward Elgar.

Ikehata, K., & Liu, Y. (2011). *Land disposal of wastes. Encyclopedia of environmental health* (pp. 353–361). Oxford: Newnes.

Intergovernmental Panel on Climate Change [IPCC]. (2007). Climate change 2007: Mitigation of climate change. In B. Metz, O. Davidson, P. Bosch, R. Dave, & L. Meyer (Eds.), *Contribution of working group III to the fourth assessment report of the intergovernmental panel on climate change*. Cambridge: Cambridge University Press.

Jaccard, M., Failing, L., & Berry, T. (1997). From equipment to infrastructure: Community energy management and greenhouse gas emissions reduction. *Energy Policy, 25*(13), 1065–1074.

Jones, T. (2008). Distributed energy systems. In P. W. Newton (Ed.), *Transitions: Pathways towards sustainable urban development in Australia* (pp. 411–424). Melbourne: CSIRO Publ.

Kennedy, C., & Corfee-Morlot, J. (2013). Past performance and future needs for low carbon climate resilient infrastructure—An investment perspective. *Energy Policy, 59*, 773–783.

Kenway, S. J., Priestly, A. Cook, S. Seo, S. Inman, M. Gregory, A. & Hall, M. (2008) *Energy use in the provision and consumption of urban water in Australia and New Zealand*. CSIRO: Water for a Healthy Country Flagship.

Lombardi, P., Carnevale, E., & Corti, A. (2011). Analysis of energy recovery potential using innovative technologies of waste gasification. *Waste Management, 32*(4), 640–652.

Lou, X. F., & Nair, J. (2009). The impact of landfilling and composting on greenhouse gas emissions—A review. *Bioresource Technology, 100*, 3792–3798.

Medeazza, G., & Moreau, V. (2007). Modelling of water-energy systems: The case of desalination. *Energy, 32*, 1024–1031.

Mountouris, A., Voutsas, E., & Tassios, D. (2006). Solid waste plasma gasification: Equilibrium model development and exergy analysis. *Energy Conversion and Management, 47*, 1723–1737.

Naess, P. (1995). Central dimensions in a sustainable urban development. *Sustainable Development, 3*, 120–129.

Newman, P. (2015). Much more than the Wild West: The rise of a sustainable city. *Griffith Review*. 47 (Late January Edition).

Newman, P., & Jennings, I. (2008). *Cities as Sustainable ecosystems: Principles and practices*. Washington DC: Island Press.

Newman, P., & Kenworthy, J. (1999). *Sustainability and cities: Overcoming automobile dependence*. Washington DC: Island Press.

Newman, P., Beatley, T., & Boyer, H. (2009). *Resilient cities: Responding to peak oil and climate change*. Washington DC: Island Press.

Newton, P. W. (2012). Liveable and sustainable? Socio-technical challenges for twenty-first-century cities. *Journal of Urban Technology, 19*(1), 81–102.

Onovwiona, H. I., & Ugursal, V. I. (2006). Residential cogeneration systems: Review of the current technology. *Renewable and Sustainable Energy Reviews, 10*, 389–431.

Pehnt, M. (2008). Environmental impacts of distributed energy systems—The case of micro cogeneration. *Environmental Science and Policy, 11*, 25–37.

PMSEIC. (2010). *Challenges at energy-water-carbon intersections*. Canberra, Australia: Prime Minister's Science, Engineering and Innovation Council.

Productivity Commission. (2010). *Population and migration: Understanding the numbers. Research paper.* Retrieved from: http://www.pc.gov.au/__data/assets/pdf_file/0016/104128/population-migration.pdf.

Rauland, V., & Newman, P. (2011). *Decarbonising Australian cities: A new model for creating low carbon, resilient cities*. Paper presented at the MODSIM2011: 19th International Congress on Modelling and Simulation, December 12–16, 2011 Perth, WA.

Reffold, E., Leighton, F. Choudhury, F. & Rayner, P.S. (2008). *Greenhouse gas emissions of water supply and demand management options*. Science Report—SC070010. Bristol.

Roseland, M. (2012). *Toward sustainable communities: Solutions for citizens and their governments* (4th ed.). Gabriola Island, BC: New Society Publishers.

Rothhausen, S. G. S. A., & Conway, D. (2011). Greenhouse-gas emissions from energy use in the water sector. *Nature Climate Change, 1*, 210–219.

Ryan, C. (2013). Eco-acupuncture: Designing and facilitating pathways for urban transformation, for a resilient low-carbon future. *Journal of Cleaner Production, 50*, 189–199.

Seadon, J. (2010). Sustainable waste management systems. *Journal of Cleaner Production, 18*, 1639–1651.

Siddiqi, A., & Anadon, L. D. (2011). The water–energy nexus in Middle East and North Africa. *Energy Policy, 39*, 4529–4540.

Sinclair, H. M. W. (2008). *Carbon neutral cities: The need for more sustainable development options in response to climate change.* Planning and urban development program, University of New South Wales. (PLAN 4132: Thesis Project).

Smart, A., & Aspinall, A. (2009). Water and the electricity generation industry: Implications of use. *Waterlines report series,* (18). Canberra: Australian Government National Water Commission.

Steffen, B., & Weber, C. (2013). Efficient storage capacity in power systems with thermal and renewable generation. *Energy Economics, 36*, 556–567.

Stocker, L., Newman, P., & Duggie, J. (2012). Climate change and Perth (South West Australia). In E. Blakely & A. Carbonell (Eds.), *Resilient coastal regions: Planning for climate change in the United States and Australia.* Washington DC: Lincoln Institute of Land Policy.

Tan, W. S., Hassan, M. Y., Majid, M. S., & Rahman, H. A. (2013). Optimal distributed renewable generation planning: A review of different approaches. *Renewable and Sustainable Energy Reviews, 18*, 626–645.

Tillman Lyle, J. (2004) Waste as a resource. In S. M. Wheeler, & T. Beatley (Eds.), *The sustainable urban development reader* (pp. 133–140). Routledge: New York.

Van Lier, J. B., & Lettiga, G. (1999). Appropriate technologies for effective management of industrial and domestic waste waters: The decentralised approach. *Water Science and Technology, 40*(7), 171–183.

Vandermeulen, V., Verspecht, A., Vermeire, B., Huylenbroeck, G. V., & Gellynck, X. (2011). The use of nomic valuation to create public support for green infrastructure investments in urban areas. *Landscape and Urban Planning, 103*, 198–206.

Walsh, M. (2011). *Cockburn coast green infrastructure study—Implementation strategy.* Industry report. Parsons Brinckerhoff, Curtin University Sustainability Policy Institute.

Went, A., Newman, P. & James, W. (2008). *Renewable transport: How renewable energy and electric vehicles using vehicle to grid technology can make carbon free urban development.* CUSP discussion paper 2008/1. Curtin University Sustainability Policy Institute. Retrieved from http://sustainability.curtin.edu.au/local/docs/cusp_discussion_paper.pdf.

Wissner, M. (2011). The smart grid—A saucerful of secrets? *Applied Energy, 88*, 2509–2518.

Wolman, A. (1965). The metabolism of cities. *Scientific American, 213*(3), 179–190.

World Business Council on Sustainable Development [WBCSD]. (2009). *Transforming the market: Energy efficiency in buildings.* Retrieved from http://www.wbcsd.org/transformingthemarketeeb.aspx.

Chapter 5
The Precinct—The New Scale for Decarbonising

> The answer is the precinct. A smart building precinct is containable and do-able, in terms of regulations, utilities and physical infrastructure. And, with luck, it will sit in one metropolitan authority.
>
> (Barker 2011: 78).

5.1 Introduction

Cities around the world have long been leading the way in acting on climate change, thus demonstrating the effectiveness of addressing carbon emissions at the local level (ICLEI 2010; Habitat 2011). However, as Choguill (2008) states, 'no single city can contribute to overall sustainability if its own component parts are found not to be sustainable' (42). These component parts of a city are its neighbourhoods and precincts. They are what a city is built around in terms of their urban design, provision of infrastructure and community networks (Bunning et al. 2013; Ryan 2013).

The neighbourhood level has been identified and discussed in the literature as a critical element for urban planning for over a century (Brody 2009; Patricios 2002). While the Garden City movement (Howard 1902) may have initially highlighted the significance of this scale, Clarence Perry (1929) is largely acknowledged as being the first to officially introduce the concept, and the importance, of the 'neighbourhood unit'. Since then, scores of people have discussed, redefined, reinterpreted and further developed the concept, continually validating the benefits of this scale. Among the more prominent thinkers were Mumford (1954), architects Stein

V. Rauland and P. Newman, *Decarbonising Cities*, Green Energy and Technology, DOI 10.1007/978-3-319-15506-7_5

(1951) and Henry Wright who developed the 'Radburn Model' (Banarjee and Baer 1984), and more recently, the work of New Urbanists[1] (Brody 2009).

Brody (2009), who provides a useful overview of the 'neighbourhood unit' literature, highlights how the concept has changed and evolved over the years as it was applied to various fields including planning, architecture, urban design, real estate, finance, law, education, transportation, public health and policing.

Despite the evolving nature of the concept, a century of the literature has proven the value of this smaller scale. It is perhaps unsurprising, therefore, that it is at this neighbourhood level where most sustainability and carbon reduction opportunities are now also presenting themselves. The abatement opportunities, together with the effectiveness of implementation, make the precinct level the optimum scale for addressing emissions within cities.

Recognition of the effectiveness of the precinct level is growing and is particularly evident when looking at building assessment tools and rating schemes. While efforts to reduce emissions from the built environment previously focused on individual buildings, as can be seen by the plethora of rating tools that exist to assess and encourage better building performance, a shift in focus has occurred in recent years to capture additional elements at the neighbourhood and precinct level (AILA 2010). Such tools have demonstrated the larger emissions abatement potential when these bigger systems within the built environment are included, such as precinct-scale energy, water, waste systems and transport options (Ewing et al. 2008; The Climate Group 2010). Buildings are thus no longer being seen in isolation, but as a part of an integrated and dependent wider precinct system.

This chapter begins by defining what we mean by 'precinct'. We highlight some of the key reasons why the precinct level is increasingly being identified as the optimal level for decarbonising. Decentralised approaches to resource management (i.e. energy, water and waste) are compared to the centralised approaches that currently dominate our cities, as the smaller, decentralised scale appears to offer numerous efficiencies.

[1] The New Urbanism movement, which has gained prominence over the years, was largely a reaction against the car-based, urban sprawl, which has developed rapidly and consistently since the 1950s and 1960s, particularly in the USA, Canada, the UK and Australia. New Urbanism sought to bring back the town centre, which included a more walkable, mixed-use urban design, allowing multiple transport modes (i.e. bike, public transit and car) and emphasising a stronger sense of community (Roseland 2005). New Urbanism has many supporters as well as critics. See McCann (2009) for a detailed analysis. Falconer et al. (2010) provides an analysis of many New Urbanist suburbs in Australia, which have not been able to live up to their ambitious goals of carbon reduction in transport.

5.2 Defining the Precinct

A precinct is loosely defined in this research as an area within a city made up of two or more buildings within a street, up to the size of a suburban development. Importantly, a precinct uses shared infrastructure, such as roads, energy, water and waste management systems. It can be a new development or a redevelopment and, while it can be purely residential or commercial; ideally, it will incorporate mixed uses, thus providing a hub or agglomeration of activities and multiple participants.

Cutler (2009: 3) proposes that a successful precinct is one that:

- Revolves around a clear core development proposition which can engage the sustained participation of multiple parties who have an expectation of and commitment to mutually beneficial interests;
- Has a sound and sustainable funding model;
- Develops a distinct identity;
- Mobilises the supply of supportive infrastructure, facilities and resources;
- Leverages individual participation and investment;
- Delivers sustained impact and benefits;
- Facilitates active collaborations and informal networking; and
- Operates with clear and transparent rules of engagement.

Interestingly, these points largely relate to issues of governance, which will be critical to the success and ultimate uptake of low carbon precincts (Bunning et al. 2013).

Precinct, community, neighbourhood and settlement are terms that are all commonly used within the literature to denote the focussed urban area with the above characteristics, and they are thus used interchangeably within this book.

5.3 Why the Precinct Scale?

There are many advantages of moving beyond the individual building scale to a precinct scale when attempting to decarbonise the built environment. This research identifies five key reasons for targeting the precinct level, as illustrated in Fig. 5.1. Firstly, by going beyond the building scale, additional elements that contribute to emissions can be factored into carbon-related decisions. Secondly, it is the scale at which communities function. Thirdly, it is the scale at which developers work. Fourthly, this level has greatest interaction with local government, and finally, it is the scale at which low carbon and emerging technologies appear to perform best. These are discussed further below.

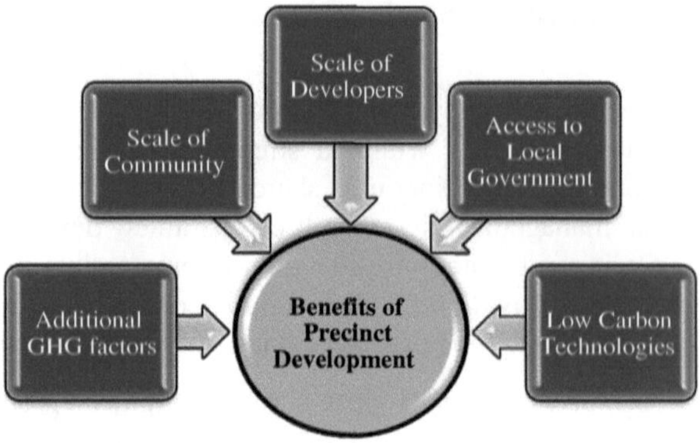

Fig. 5.1 Key reasons for targeting the precinct scale for emissions abatement

5.3.1 Inclusion of Additional Urban Factors

When discussing the built environment in terms of sustainability and its emissions contribution, the traditional focus has predominantly been on buildings. However, broadening the Scope to the precinct scale allows a variety of additional factors to be taken into consideration, many of which contribute significantly to urban carbon emissions but are often seen as beyond a developer's immediate control. By incorporating these additional factors, such as the transport options within a precinct and its management of resources, a deeper understanding can be gained of the total carbon implications associated with the built environment.

A framework that allows a developer to assess and compare various design approaches based on their costs and benefits can enable developers to make more informed decisions on important precinct infrastructure (Newman et al. 2011). This will help to lower the overall footprint of the development and thus make the process of becoming carbon neutral significantly easier.

5.3.2 Scale of Communities

The social and cultural dimensions of cities have been well recognised as a key factor in achieving sustainability outcomes (Choguill 2008; Ryan 2013). The social connections and interaction between people largely occur at the community scale. Middlemiss and Parrish (2010) note that the 'community' is increasingly identified 'as a space for realising pro-environmental change' (7560). As a result, it is receiving greater attention by policy makers and academics, evidenced by the increasing number of behaviour change initiatives targeting and engaging the community on climate change and sustainability issues.

The community has also been identified as a place where organic social change occurs in the form of the grass roots initiatives. Seyfang (2010) identifies communities as innovative niches with the potential for wider societal transformation largely due to the local networks, businesses and organisations that exist at the community level. The unique set of skills, experience and local knowledge that exists within local communities puts them in an ideal position to respond to local environmental challenges and issues more effectively (Habitat 2011).

Communities also play an important role in shaping social norms and practices (Fuller 2009). Moloney et al. (2010) recognise that 'normalised pro-environmental behaviour' (7622) is a key factor in transitioning to low carbon communities. While having the right technical and infrastructural solutions in place is important to provide opportunities for people to make the right choices, societal transformation 'has to include the interaction between social and cultural life and physical and technical systems' (Ryan 2013: 193).

5.3.3 Scale of Developers

Precincts and neighbourhoods are the scale at which development generally occurs (Sharifi and Murayama 2013), and therefore, the scale at which developers are used to working. Developers are also used to interacting with local governments on regulatory and planning issues, which can create opportunities for greater collaboration and working together to share both the costs and benefits associated with creating low carbon communities.

5.3.4 Access to Local Government

Local government has long been identified as an ideal scale for tackling climate change. The development of Local Agenda 21 and ICLEI—Local Governments for Sustainability, along with the introduction of various sustainability initiatives such as cities for climate protection (CCP), are evidence of the willingness and effectiveness of local councils in addressing sustainability and climate change issues (Broto and Bulkeley 2013; Brugmann 1996, 2010; ICLEI 2010; World Bank 2010).

Since local governments have authority over many of the planning controls within their local government areas, such as land-use zoning, buildings, waste management and other common infrastructure (World Bank 2010), they are well positioned to create strategic policies for developers that ensure certain sustainability outcomes. Furthermore, as local governments already manage many of the systems and infrastructure and are increasingly being seen as 'early innovators and adopters of efficient technologies' (Burch 2010: 7575), this can provide an excellent basis for developers in creating low carbon developments.

Local government is not only the closest level of government to developers, but also to the community. They therefore have the greatest ability to influence individuals and affect change within the community, by enabling better dialogue between citizens, developers and local decisions-makers (World Bank 2010).

5.3.5 Scale of Emerging Low Carbon Technologies

Precincts provide an ideal platform to trial and deploy existing technologies and systems in an integrated manner to achieve the greatest environmental, economic and social outcomes. They provide the scale to test commercial viability and the ability to evaluate and replicate solutions. Working at the level of the precinct affords an opportunity to test the 'systems thinking' needed to better understand the many and varied interactions between the infrastructure and systems that underpin urban life.

The precinct is increasingly being identified as an ideal scale for new and emerging low carbon technologies (The Climate Group 2010). Examples of the efficiency gains available through localised systems include the use of co- and tri-generation, renewables, localised water systems, solid waste recycling systems and many technologies associated with resource efficiencies, which were discussed in the previous chapter. This scale provides great opportunities to test multiple technologies and systems in an integrated fashion at this smaller city/precinct level. Integrating technologies, systems and planning in a strategic way can provide numerous economic and environmental benefits and efficiencies that can be 'an order of magnitude greater than when they are pursued in isolation' (The Climate Group 2010: 9).

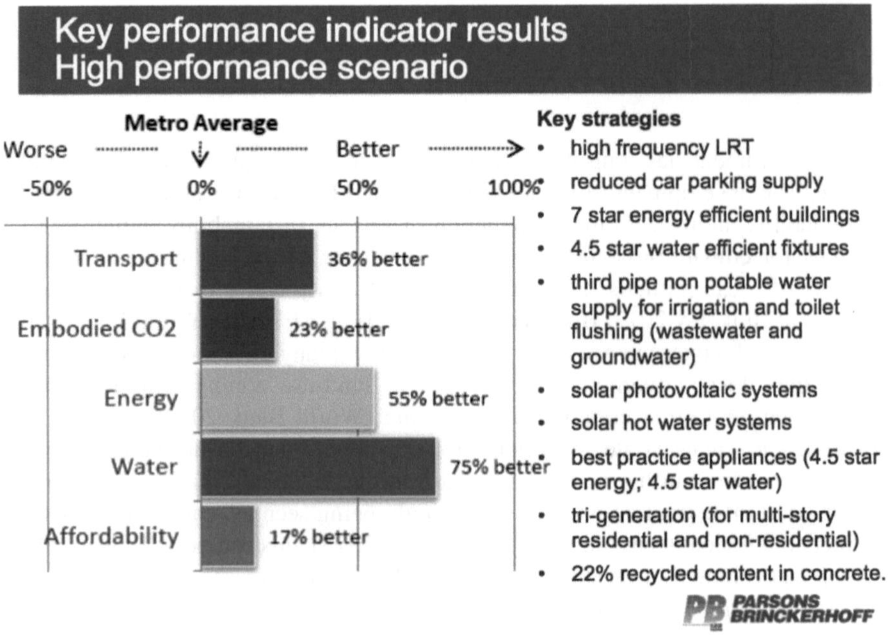

Fig. 5.2 Key performance results from integrated low carbon infrastructure in the Cockburn Coast. *Source* Walsh (2011)

The precinct scale can allow new business cases and models to be developed and trialled, many of which are or will be far more viable than at larger scales. Figure 5.2 below demonstrates the financial and environmental savings generated from various technologies and infrastructural measures implemented at the precinct level on the Cockburn Coast project described in Chap. 4.

Targeting the precinct level helps not only to mitigate climate change by offering significant emissions reduction opportunities, but also to build resilience in terms of adapting both to climate change impacts and to broad resource availability issues. Communities that are able to meet more of their resource needs from their local environment will decrease their risk of global resource and commodity price shocks.

5.4 Conclusion

While the importance of 'the neighbourhood' and precinct have long been recognised and discussed for their role in the design of cities, this scale is now also being recognised for its ability to achieve carbon reduction within the built environment. Moving beyond the individual building to the precinct enables a range of new technologies to become viable and a mix of important social interactions to occur, which all help to deliver high performance, low carbon urban development. The next chapter highlights some eminent examples of this type of development and evaluates them in terms of their emissions' sources targeted.

References

Australian Institute of Landscape Architects [AILA]. (2010). *Climate adaptation tools for sustainable settlements (CATSS)*. Canberra, Australia: Australian Institute of Landscape Architects.

Banarjee, T., & Baer, W. C. (1984). *Beyond the neighborhood unit: Residential environments and public policy*. New York: Plenum Press.

Barker, T. (2011). *A book of wicked problems—A compendium of projects for the city and regional centres*. Broadway, NSW: UrbanAid.

Brody, J. (2009). *Constructing professional knowledge: The neighborhood unit concept in the community builders handbook*. Doctor of Philosophy Dissertation, University of Illinois at Urbana-Champaign, Urbana, Illinois.

Broto, V. C., & Bulkeley, H. (2013). A survey of urban climate change experiments in 100 cities. *Global Environmental Change, 23*, 92–102.

Brugmann, J. (1996). Planning for sustainability at the local government level. *Environmental Impact Assessment Review, 16*(4–6), 363–379.

Brugmann, J. (2010). *Welcome to the urban revolution: How cities are changing the world*. New York: Bloomsbury Press.

Bunning, J., Beattie, C., Rauland, V., & Newman, P. (2013). Low-carbon sustainable precincts: An Australian perspective. *Sustainability, 5*(6), 2305–2326.

Burch, S. (2010). In pursuit of resilient, low carbon communities: An examination of barriers to action in three Canadian cities. *Energy Policy, 38*, 7575–7585.

Choguill, C. (2008). Developing sustainable neighbourhoods. *Habitat International, 32*, 41–48.

Cutler, T. (2009). The role of precincts in innovation systems. There's no place like a precinct…an overview. Retrieved from http://www.cutlerco.com.au/activities/speeches/09_speeches/precincts_and_innovation_overview.pdf.

Ewing, R., Bartholomew, K., Winkelman, S., Walters, J., & Chen, D. (2008). *Growing cooler: The evidence on urban development and climate change*. Washington, DC: Urban Land Institute.

Falconer, R., Newman, P., & Giles-Corti, B. (2010). Is practice aligned with the principles? Implementing New Urbanism in Perth, Western Australia. *Transport Policy, 17*(5), 287–294.

Fuller, J. (2009). *Heads, you die: Bad decisions, choice architecture, and how to mitigate predictable irrationality*. Surrey Hills, NSW: Per Capita Australia Limited.

Habitat, U. N. (2011). *Cities and climate change: Global report on human settlements 2011*. London: Earthscan.

Howard, E. (1902). *Garden cities of tomorrow*. London: S. Sonnenschein & Co., Ltd.

ICLEI. (2010). Cities in a post-2012 climate policy framework. *ICLEI Global Reports*. Retrieved from http://ccsl.iccip.net/cities_in_a_post_2012_policy_framework.pdf.

McCann, E. J. (2009). New Urbanism. *International Encyclopedia of Human Geography, 7*, 438–443.

Middlemiss, L., & Parrish, B. D. (2010). Building capacity for low-carbon communities: The role of grassroots initiatives. *Energy Policy, 38*, 7559–7566.

Moloney, S. E., Horne, R., & Fien, J. (2010). Transitioning to low carbon communities—From behaviour change to systemic change: Lessons from Australia. *Energy Policy, 38*, 7614–7623.

Mumford, L. (1954). The neighborhood and the neighborhood unit. *Town Planning Review, 24*, 256–270.

Newman, P., Rauland, V. & Holden, D. (2011). Creating resilient cities: How a new generation of tools can assist local governments in achieving their carbon abatement goals. 5th *State of Australian Cities National Conference, Nov 29–Dec 2 2011*, Melbourne, Vic.

Patricios, N. (2002). Urban design principles of the original neighborhood concepts. *Urban Morphology, 6*(1), 21–32.

Perry, C. A. (1929) 'The neighbourhood unit', monograph 1 in committee on regional plan of New York and its environs (ed.) Neighbourhood and community planning. Regional survey volume VII (committee on regional plan of New York and its environs, New York), 20–141.

Roseland, M. (2005). *Toward sustainable communities: Resources for citizens and their governments*. Canada: New Society Publishers.

Ryan, C. (2013). Eco-acupuncture: designing and facilitating pathways for urban transformation, for a resilient low-carbon future. *Journal of Cleaner Production, 50*, 189–199.

Seyfang, G. (2010). Community action for sustainable housing: Building a low-carbon future original research article. *Energy Policy, 38*(12), 7624–7633.

Sharifi, A., & Murayama, A. (2013). A critical review of seven selected neighborhood sustainability assessment tools. *Environmental Impact Assessment Review, 38*, 73–87.

Stein, C. S. (1951). *Towards New Towns for America*. Chicago: University Press of Liverpool.

The Climate Group. (2010). *Climate smart precincts: Opportunities report to the first ministers*. London, UK: The Climate Group.

Walsh, M. (2011). *Cockburn coast green infrastructure study—Implementation strategy*. Industry Report. Parsons Brinckerhoff, Curtin University Sustainability Policy Institute.

World Bank. (2010). *Cities and climate change: An urgent agenda* (Vol. 10). Washington DC: Urban Development & Local Government.

Chapter 6
Eco-Precincts

6.1 Introduction

The concept of an 'eco-city' has its roots in the early twentieth century (Roseland 1997). However, interest in the concept grew rapidly in the late 1980s when people began exploring how the newly defined notion of 'sustainable development' (World Commission on Environment and Development 1987) could be interpreted within an urban context and at a city scale. Numerous articles began investigating what eco-cities may look like and the many elements that should be incorporated into their design and management. These writings were instrumental in helping to popularise the concept (Roseland 1997).

Fast-forward two decades, and the previously aspiring eco-city concept[1] (Engwicht 1992) has become a reality with countless demonstrations around the world. An eco-city survey conducted by the University of Westminster identified over 170 eco-cities globally (Joss et al. 2011: 135). The cities range in size and scale, as well as in ambition. Several have drawn considerable attention over the years due to their progressive nature, inspiring and impressive goals and successful realisation. Best known examples of these include BedZED in the UK, Vauban in Freiburg, Germany, and Hammerby Sjostad in Stockholm, Sweden.

This process of rapid adoption of eco-city innovations and models has all the marks of a social movement (Roseland 1997; Snow et al. 2004). The importance of eco-cities being a social movement is that they are likely to grow from a broad base and influence governments to facilitate them further in their planning, governance and, hopefully, certification processes.

Several factors are likely to have contributed to this rise in eco-cities. For some cities, the threat of climate change and their unique vulnerability to the varying impacts of it are enough motivation to take action. For others, it may be a response to the lack of action being taken at higher levels of government, and the

[1] Terms such as "towards" were often used in describing the development of eco-cities, implying that they were still only an aspirational concept.

V. Rauland and P. Newman, *Decarbonising Cities*, Green Energy and Technology, DOI 10.1007/978-3-319-15506-7_6

encouragement and empowerment provided by international initiatives such as ICLEI (Newman et al. 2009). For many cities, however, it is simply a chance to differentiate and showcase themselves in an increasingly ambitious and competitive world. Whatever the reason is, larger and grander environmental targets and goals for cities and districts are constantly being set.

Unsurprisingly, various new terms have emerged from this highly diverse set of cities and precincts to help describe what they are seeking: carbon zero (or zero carbon), zero emission, carbon free, low carbon, carbon negative, carbon positive, climate positive, and climate neutral. However, the myriad of names are rarely well defined, and the methodology used for calculating emissions associated with the developments, or the process by which a development has achieved its stated carbon goal, often remains unclear. This lack of consistency in information has led not only to confusion, but also to scepticism and mistrust in claims. Environmental claims that are considered misleading or false are commonly characterised by the term 'greenwash' and have become increasingly prevalent (Monbiot 2006).

The purpose of this chapter is to investigate several of the more widely known and celebrated eco- and low-carbon cities and developments around the world, in order to gain a better understanding of the specific areas (e.g., energy, water, waste, transport) that have been targeted and how each area is reflected in their eco- or carbon claim. Focus is given to carbon emissions, as this is a growing area of concern and thus increasingly exhibited in claims.

The developments analysed include the following: BedZed in the UK; Vauban in Germany; Hammerby Sjostad in Stockholm and Western Harbour in Malmo, Sweden; Dongtan in China; and Masdar City in the United Arab Emirates. In Australia, two municipalities—the City of Sydney and City of Fremantle—have claimed carbon neutrality, and a proposed new development off the coast of Fremantle (North Port Quay) claims to be carbon free. A brief analysis and comparison of these nine cities and precinct developments is provided below.

6.2 International Case Studies

6.2.1 BedZED—UK

BedZED is a small, dense, mixed-use eco-development located in the southern suburbs of London (see Fig. 6.1). It is built on a brownfield site and situated close to a rail station. From its inception, it pushed the boundaries, testing and showcasing several new low-carbon technologies and urban design elements. Completed in 2002, BedZED is one of the original and hence most well-known global examples of a grass roots sustainable development.

Fig. 6.1 BedZED. *Credit* Jan Scheurer

6.2.1.1 Eco-Claim

The name BedZED originates from Beddington Zero (Fossil) Energy Development and was originally designed to be a carbon-neutral precinct (Hodge and Haltrecht 2009). The aim was to generate enough zero-carbon energy onsite to meet the electricity, heating and hot water needs of the development, while feeding excess energy into the grid.

6.2.1.2 Low-Carbon Initiatives

Despite an overarching focus on operational energy, the designers of the development integrated a variety of sustainability measures and initiatives that targeted emissions from a wide range of areas, including water, waste, materials and transport. The development's broad focus on sustainability made it an outstanding example of what is possible at a small-scale and very local level. Some of the initial low carbon, sustainability features included the following:

- highly energy-efficient three-story townhouses that use solar passive design, including passive ventilation and daylighting to reduce the need for artificial heating, cooling and lighting;
- low-carbon materials sourced locally where possible;
- energy- and water-efficient appliances;
- grey water-recycling facilities onsite as well as rainwater tanks;
- solar photovoltaic panels on roofs;
- biomass-fuelled, combined heat and power (CHP) plant providing zero-carbon electricity and district heating to the development (later changing to a gas district-heating system);
- mixed use, combining office space and residential dwellings, to reduce transport emissions from commuting;
- transit-oriented (i.e. situated next to a train station to promote low-carbon transportation) to reduce personal transport emissions; and
- provision and encouragement of local food production (including rooftop gardens), reducing food miles and associated emissions.

In addition to the numerous low-carbon initiatives, many other broad sustainability features were included such as affordable housing and a range of other housing tenures (Williams 2012), community facilities and open space, and the preservation and promotion of biodiversity in the surrounding area (Newman et al. 2009; Newman and Jennings 2008).

6.2.1.3 Success

BedZED continues to be one of the most widely documented and celebrated case studies demonstrating successful, holistic, and sustainable precinct design. However, as Hodge and Haltrecht (2009) point out, despite thousands of articles and reports written about the development,[2] only a few contain performance-related information. From the information available, the main issue BedZED experienced concerned the malfunctioning of the biomass plant (Williams 2012), in which the development relied upon to provide the zero-carbon energy to the precinct, and hence to support the development's name.

Since the biomass plant ceased operation in 2005, electricity has been sourced primarily from the national (i.e. coal-fired power) grid, with a small contribution coming from onsite solar photovoltaic panels. A new district-heating system using natural gas now supplies heating (Williams 2012) that is far cleaner than electricity from the grid. This, together with the highly energy-efficient buildings and appliances, means the amount of grid electricity consumed by the development is around half of that of the neighbouring areas (Hodge and Haltrecht 2009).

[2] A Google search of 'BedZED' on the 26/03/12 listed 45,200 results.

Nevertheless, the use of grid electricity means that carbon emissions are now associated with the development, demonstrating the potential legal implications that could arise from what now appears to be a false carbon claim.

The initial wastewater treatment system was also decommissioned early on, although a new MBR reactor has since taken its place. The reason provided for the initial decommissioning was unsustainable operating and maintenance costs. It is not yet known how the newer technology will fare in comparison, though it is expected that rising water costs associated with climate variability and water scarcity will make this technology more commercially viable in the near future (Hodge and Haltrecht 2009).

The impact of this system on the development's carbon emissions needs to be investigated, however, since the energy required to run small-scale wastewater systems can be greater than the energy needed in conventional municipal wastewater treatment plants (Medeazza and Moreau 2007). If energy required to operate the system is not sourced from renewables, it will result in higher emissions.

Despite the challenges and issues experienced, BedZED has demonstrated that massive carbon abatement opportunities are possible at the precinct scale and continue to be a shining example for other developments.

6.2.2 Vauban, Freiburg—Germany

Vauban is another relatively compact, sustainable neighbourhood development located on the southern edge of the city of Freiburg, in Germany. Redeveloped on a former military base, Vauban is largely residential, designed to meet rising housing demand (Williams 2012). It is connected by a light rail, providing residents with a short commute to the city centre and is virtually car free with limited parking provided at each house and full walking and cycling orientation.

The development is renowned for its passive and active integrative solar design onsite, biomass co-generation plant, integration of nature (Fig. 6.2) and car-free or shared-zone streets (Fig. 6.3) (Newman et al. 2009; Newman and Jennings 2008; Williams 2012). The houses are also surrounded by plants and gardens designed and built between 1995 and 2008, with considerable community consultation; this development demonstrates what is possible when citizens and residents are involved and empowered through the planning process.

6.2.2.1 Eco-Claim

Vauban is often referred to as an eco-village, sustainable urban district or 'model sustainable neighbourhood' (Newman and Jennings 2008). While it does not make any specific carbon claim, studies have shown that Vauban has reduced carbon

Fig. 6.2 Apartments in Vauban. *Credit* Vanessa Rauland

Fig. 6.3 A mixed-use street in Vauban. *Credit* Peter Newman

within its district considerably. Williams (2012) notes that Vauban residents produce around 0.5 tonnes CO_2/per person/year compared to the average 8.5 tonnes CO_2/per person/year for typical Freiburg residents. The list of initiatives below offers insight into how they were able to achieve this.

6.2.2.2 Low-Carbon Initiatives

Vauban's low-carbon features include the following:

- energy-efficient houses (with all buildings meeting at least the low-energy standard, though 200 are passive houses/units and 59 Energy-Plus houses);
- expansive solar photovoltaic systems on roofs of houses, public and commercial buildings and car parks;
- biomass-fuelled co-generation plant (together with solar supplying 25 % of districts electricity needs);
- heavily restricted car parking facilities deterring car ownership (with only 16 % of Vauban residents owning a car);
- a mix of car-free and shared-zone streets (with the majority of travel being by foot or bike);
- light-rail connection to the city;
- adoption of housing cooperatives[3] and associations, which allows residents to have responsibility for the management of their building, and thereby increases their awareness around issues such as energy consumption; and
- mixed-use development (around 600 jobs and 5000 residents).

While it is unclear which measures were included in the carbon footprint, it is likely that the low-per capita footprint would be largely attributable to the buildings in Vauban—all of which meet high energy efficiency standards. Although the houses are extremely energy efficient from an operational perspective, it is unclear whether the embodied emissions in materials used in the construction of the buildings have been accounted for.[4]

Another reason for the low-carbon footprint in Vauban would be related to the energy system supplying the precinct's electricity. Described by Williams (2012) as low or zero-energy system (LZE), co-generation using biomass and solar photovoltaics supplies one quarter of the district's electricity.

6.2.2.3 Success

Vauban adopted a very different approach to developing this community, which has largely contributed to its success. As Williams (2012) notes, 'Vauban was conceptualised through a highly participative planning process with residents, planners, housing companies, Baugemeinschaften/Baugruppen [housing co-ops and associations], utilities all working together to achieve a sustainable neighbourhood' (139). This collaboration became a critical factor, particularly in helping to deliver such

---

[3] Co-operative housing is a situation where citizens join together to design, build and own a shared housing estate—often in the form of an apartment block.

[4] Indeed, high operational efficiency often requires more energy intensive materials such as concrete or bricks.

high energy efficiency standards, as the higher upfront costs associated with many of the initiatives were more accepted and supported by the residents and locals who, having a long-term vested interest in the development, could clearly see the benefit of the energy savings and lower overall running costs. This approach provides a great model for overcoming the problem of split incentives (discussed in more depth in Chap. 13).

Allowing residents to co-design and co-manage their buildings, energy systems and community infrastructure has led to greater sense of responsibility for and ownership of the overall success of the sustainability initiatives, and this sets it apart from most other developments. While it provides a brilliant and important example and model for other developments, it is worth noting that such a development relies on the 'greenness'[5] of residents and their willingness to be active in the decision-making process.

6.2.3 B001, Western Harbour, Malmo—Sweden

B001 is an eco-development built in the Western Harbour district of Malmo in Southern Sweden. It is a mixed-use, high-density precinct, currently home to 3000 residents as well as some light rail, commercial and tourism businesses. By 2020, the development is anticipated to house between 15,000 and 20,000 residents and attract several eco-businesses to the district (The Oresund Committee 2009; Williams 2012).

The development is positioned in a prime location on the Harbour (see Fig. 6.4), a 10-min bus ride to the town centre of Malmo and a 20-min train ride to Copenhagen from there. It was previously an industrial precinct and thus in need of decontamination renewal. It is now a popular and attractive residential and tourist location.

6.2.3.1 Eco-Claim

B001 is most often cited as an eco-district, but has also been labelled as carbon neutral, climate neutral and energy neutral (The Oresund Committee 2009). Interestingly, there appears to be little information about, or comparison of, the residents' carbon footprints, which seems to be a common metric used by other eco-developments. Most literature pertaining to B001 discusses the various initiatives in place, particularly focussing on the energy efficiency standards of the buildings and the low-carbon energy system employed (Williams 2012).

[5] Willams (2012) notes that most residents that chose to move to Vauban had ethical concerns for the environment and around 70 % vote Green.

Fig. 6.4 B001, Malmo, Sweden. *Credit* Vanessa Rauland

6.2.3.2 Low-Carbon Initiatives

Some of the low-carbon initiatives implemented at B001 include the following:

- 100 % locally generated sustainable energy;
- a wind turbine in the harbour, 3 km from the development, supplying energy to houses;
- solar PV embedded into the development;
- green roofs;
- heating/cooling systems using a local aquifer as storage;
- high energy efficiency standards of buildings (105 kW/m^2); and
- a waste to energy system.

6.2.3.3 Success

The design and development of the precinct, including the low-carbon energy system supporting it, were achieved through a highly collaborative process involving the municipality and its specially created 'sustainability unit', together with planners, developers and a local utility company. Unlike Vauban, B001 did not engage the community in this process.

The residents who ultimately moved into the area were relatively affluent and less concerned or aware of the environmental status of the development, but they chose to live there based on the aesthetic quality and location of the development. This most likely contributed to the higher-than-expected energy consumption of the buildings within the development (Nilsson and Elmroth 2005). Whether this was the fault of the developers not meeting the correct energy efficiency standards, or of the modelling not accounting for the affluent people moving into the district, the high energy standards (105 kWh/m^2) set for the first phase of the development were not met.

Nevertheless, the low-carbon energy supply system for B001 has experienced relative success, and despite the overconsumption of energy, the fact that the districts' energy comes from low-carbon sources ensures that the carbon footprint remains low. What the development's carbon footprint actually is remains to be seen, as little research is available. A drawback from the energy system, as Williams (2012) notes, is that it allows residents to be passive consumers (unlike in Vauban where they have a more active role in energy management), which is most likely leading to the higher energy usage.

Being municipality-owned land, the process of converting Western Harbour to an eco-development and enforcing environmental standards was made much easier than if the site had been privately owned. However, as demonstrated by post-occupancy evaluation (Nilsson and Elmroth 2005; Williams 2012), such enforcement does not always ensure delivery of the desired outcome.

6.2.4 Hammarby Sjostad, Stockholm—Sweden

Hammarby Sjostad is a 200-ha site located in a close proximity to the city centre of Stockholm. Previously contaminated municipality-owned land with poor transport connectivity, Hammarby Sjostad has now become a vibrant, desirable and celebrated eco-district within Stockholm (see Fig. 6.5). The district houses 20,000 residents in a relatively dense area (50 units/ha) and boasts one of the most celebrated and discussed 'closed-loop' systems for integrating energy, water, waste and transport at the local level. This has contributed considerably to the reduction in the district's emissions, and Hammarby's prime location (waterfront and close to the city) has been optimised by the extension of a light rail to the development.

Fig. 6.5 Hammerby Sjostadt, Stockholm, Sweden. *Credit* Vanessa Rauland

6.2.4.1 Eco-Claim

Hammarby Sjostad is well known as an 'eco-district' with its eco-claim predominantly centred around its innovative eco-cycle model or closed-loop resource management system. The development sets the goal of reducing per capita emissions within the development to 2.5–3 tonnes CO_2/per person/year compared with the average Stockholm citizen's emissions of 4 tonnes CO_2/per person/year (Williams 2012).

6.2.4.2 Low-Carbon Initiatives

Some of the low-carbon initiatives implemented in Hammarby Sjostad include the following:

- a closed-loop system that utilises the waste products from various processes such as waste water and municipal solid waste to create energy (biofuel, biogas and waste heat), which is used for electricity generation, district heating and fuel in cars;
- micro-generation, i.e., from solar PVs and solar collectors;
- energy-efficient buildings;
- vacuum waste collection;
- light rail connected to city;
- a car share scheme; and
- an Environmental Advice Centre to inform and educate residents about living a low-carbon lifestyle.

(*Sources* Newman et al. 2009; Takahashi and Onishi 2011; Williams 2012).

As Stockholm residents already have one of the lowest per capita carbon footprints for a developed city (largely because of the low-carbon energy grid, district-heating network supported by relatively high density, and thermally efficient buildings), it was difficult to find and implement further reduction measures.

6.2.4.3 Success

The design and development of the closed-loop system came about from a collaboration between the various city-owned utilities (Stockholm Energi, Stockholm Water Company and Stockholm Waste Management Administration) and was facilitated by the municipality (Williams 2012). Sweden has a long history of municipality-owned utilities, as well as decentralised infrastructure, both of which would have contributed significantly to the initial success in setting the system up. While the system has now been privatised, it continues to operate on a profitable basis.

The entire development was planned using a highly collaborative process, which involved developers, planners, utilities and the municipality's environmental unit. Initially, strict requirements were placed on the energy performance of the buildings, which was possible because the municipality owned the land and would only permit developers who met their criteria. However, a change of government saw the land sold off to the private sector, which meant environmental standards were lowered and harder to enforce. As a result, the energy efficiency targets for buildings (50 % less energy) were not met, though there was a reduction of approximately 30 % (Williams 2012).

Despite the environmental credentials of the Hammarby Sjostad and the Environmental Advice Centre offering information and advice on low-carbon living, it appears that the residents are less concerned about environmental performance or their own footprint than are the residents of Vauban or BedZED. As was the case in B001, Williams (2012) notes that residents moved there primarily based on the 'location, aesthetic quality and services' (126), and the design of the low carbon system (very automated with little personal engagement required) allowed residents to continue as passive consumers of energy and resources.

Despite this, research suggests citizens of Hammarby produce less than half the CO_2 of other parts of Stockholm (Takahashi and Onishi 2011), though this research fails to elaborate on what emission sources are included, again making comparisons with other developments tricky.

6.2.5 Masdar—United Arab Emirates

Masdar City is located on the outskirts of Abu Dhabi in the United Arab Emirates (UAE). Situated in a desert environment with average summer temperatures of 35 degrees, limited supply of fresh water and only road access (currently),[6] it is vastly different to the other case studies. Nevertheless, it has highly ambitious aspirations to be a zero-carbon, zero-waste city. The concept, first conceived of in 2006, was designed to push the limits and provide a test bed for new technologies (Kansara and Ridley 2012).

The idea for the development stemmed from the acceptance that the source of UAE's prosperity and subsequent excessive carbon footprint, i.e., fossil fuels, would not last forever. Masdar is an attempt to build capacity and a new income stream for Abu Dhabi in a post-carbon world through attracting green economy businesses to locate there (Reiche 2009).

Once completed, Masdar is expected to house 40,000 residents and support a further 50,000 daily commuters, making it the largest low-carbon eco-city in the world to date (Reiche 2009).

[6] A train line is proposed into the future.

6.2.5.1 Eco-Claim

Masdar has had countless carbon distinctions attributed to it, with many journalists, researchers, designers and locals preaching its environmental credentials. Some of the titles include carbon neutral, net zero carbon emissions, low carbon and a zero-waste city. The development has the aim of reducing energy consumption by 80 % compared to the average Abu Dhabi resident (Kansara and Ridley 2012), and it is to be powered by 100 % renewable energy.

6.2.5.2 Low-Carbon Initiatives

Unlike the previously discussed low-carbon developments, where it was unclear as to which emissions were calculated in the carbon footprint and thus underpinned their eco-claim, Masdar is very transparent and thorough in its calculation of emissions. There are six areas that are taken into account and calculated as part of the development's carbon footprint. These are represented in Fig. 6.6.

The following initiatives all contribute to reducing the development's emissions to zero:

- highly energy-efficient buildings using at least 30-cm-thick insulation;
- traditional narrow streets for optimum shade and cooling, and funnelling of cool breezes;
- solar chimneys to draw hot air out of urban areas (see Fig. 6.7);
- absorption chillers run on solar-heated hot water to provide air conditioning;
- waterless urinals and low-flow fixtures in buildings;
- the treatment of wastewater for reuse, with the by-products used for electricity generation;
- vacuum waste collection, sorting and recycling;
- gasification of waste to produce energy, with the by-product to be used in construction (and thus zero waste);
- solar PVs supplying the development's electricity during the day;
- an electric personal rapid transit (PRT) system for transport around the city;
- light rail linking Masdar to Abu Dhabi; and
- ongoing data collection and monitoring of all initiatives and technologies to test for improvements and to verify carbon claims.

6.2.5.3 Success

The first phase of the development has been completed, consisting of six buildings including student accommodation that is already housing students.[7] While this could be viewed as a success, plans to develop it further (i.e. to complete the vision of

---

[7] See http://www.renewablepowernews.com/archives/2736.

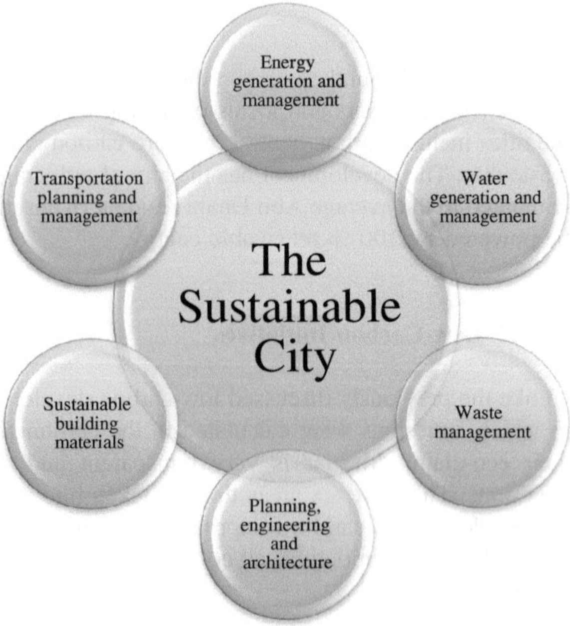

Fig. 6.6 Elements of Masdar's sustainable city concept. *Source* adapted from Masdar City (2011)

Fig. 6.7 Solar Chimney, Masdar, Abu Dhabi, UAE. *Credit* Peter Newman

accommodating 40,000 residents) have stalled due to the global financial crisis (GFC) with no further notice of resumption, questioning the overall viability of the project.

While many are impressed and inspired by what Masdar is attempting, others see Masdar as an unhelpful case study as the budget far exceeds what most ordinary cities could afford. However, it does represent an important test bed for emerging technologies and new building techniques and, with its ongoing data collection and reporting of emissions, it can help to inform new models of city development into the future.

In terms of carbon claims, Masdar provides the most comprehensive carbon analysis of all the developments—compiling data on the carbon emissions associated with every foreseeable aspect of the development including the flights and taxis of international consultants. The development is being very transparent and sets an excellent precedent, demonstrating what is possible when calculating the carbon footprint of a development.

Nevertheless, it is likely that this amount of measurement was only achievable because of the available funds. Such a rigorous and comprehensive process would no doubt be extremely time-consuming and unaffordable for most eco-developments.

6.2.6 Dongtan—China

Dongtan was to be a new sustainable city located on Chongming Island in the Yangzte River Delta, close to Shanghai. Phase one of the development was going to accommodate 25,000 people and was planned to be completed by 2010. Once finished, Dongton was anticipated to be home to more than 500,000 residents (World Bank 2009).

6.2.6.1 Eco-Claim

Dongtan had various environmental commendations bestowed upon it and eco-titles ascribed to it. It was heralded as an eco-city, zero-carbon city, sustainable city and ecologically sustainable city.

The development was credited for encouraging biodiversity, reducing car use and promoting low-consumption lifestyles, and was predicted to have less than half the carbon footprint of Shanghai's residential areas (Williams 2009).

6.2.6.2 Low-Carbon Initiatives

Initiatives were going to include the following:

- zero-energy and passive structures[8];
- recycling of 80 % of waste;

[8] For more information, see http://www.dac.dk/en/dac-cities/sustainable-cities-2/show-theme/energy/dongtan-the-worlds-first-large-scale-eco-city/?bbredirect=true.

- electricity and heat provided from renewable sources onsite, including a CHP or co-generation plant running on biofuels;
- mixed-use, medium density development (up to eight stories), with most residents expected to work within the city;
- public transport available within 7-min walk of all housing;
- emission-free vehicles; and
- locally produced food.

6.2.6.3 Success

While the planning and design of Dongtan began more than eight years ago, the only construction that has occurred on the site was the clearing and preparation of the land (which included relocating 650,000 people) and the development of a small information centre, which has also since shut down (Williams 2009). The site has lain dormant since.

There are several possible reasons for the failure of the project. While corruption is often rumoured to be a major cause of its demise, other reasons given include the overprescriptive nature of China's planning industry (progressive, innovative development required changes to rules, which is hard in the Chinese context) and political fallout (i.e. rumours of corruption lead to lack of support).[9] With global engineering firm Arup leading the design and construction, few appeared to critique the development's design, or suggest it as a potential reason for the failure.

Though, one of the few who did criticise the development was international journalist Austin Williams (2009), who suggested that the development was 'squat, low-rise, parochial, carbon-fetishising, architecturally unappealing, unworkable urban eco-clichés'. The design perhaps contributed to its demise and as Williams concluded Dongtan 'is a recipe for future disasters'.

It is reasonable to expect that problems would have ensued from the proposed densities for Dongtan, which were indeed similar to many English new towns, many of which have also turned out to be highly car dependent despite their architectural drawings which showed most people walking and cycling (Dac and Cities 2012).[10]

While a maximum of eight stories may represent high density in some European locations, the Chinese market is likely to require significant expansion and increase in density for it to be considered economically viable and representative of typical Chinese cities. A city of 500,000 (the expected population on completion) is still paltry in comparison with the megacities expected, and indeed planned, in China in the coming years. For a Chinese eco-city to be successful and replicable, it should be representative of realistic future cities, including through the application of realistic densities.

---

[9] See http://www.guardian.co.uk/environment/2009/apr/23/greenwash-dongtan-ecocity.

[10] See http://www.dac.dk/en/dac-cities/sustainable-cities-2/show-theme/energy/dongtan-the-worlds-first-large-scale-eco-city/?bbredirect=true.

Nevertheless, Dongtan was very successful in promoting and publicising the project and thus was able to put the eco-city idea and concept onto the wider agenda in China. As a result, over 100 eco- and low-carbon cities are now planned (Wu 2012).

6.3 Australian Case Studies

There are few notable demonstration projects implemented in Australia that are similar to those highlighted above. Central Park, a low carbon, sustainable highrise development in Sydney, would be one exceptional example, though it is not at the scale of the previously described case studies. For more information on this development, see http://www.centralparksydney.com/. There are two local councils that have become carbon neutral, as well as an ambitious proposed carbon-neutral island development on the coast of Western Australia. These three innovative case studies are discussed below.

6.3.1 City of Sydney—NSW

The City of Sydney encompasses a land area of 26 km^2, which includes Sydney's central business district (CBD), as well as 30 surrounding suburbs, all of which the local government authority has responsibility over (see Fig. 6.8). Although the city has a residential population of 183,616, the overall population of the district expands to over one million on a daily basis. The city is therefore responsible for a large number of operations and services to manage the large number of people living, working and visiting the city. Sydney CBD is the most important part of Australia's economic infrastructure in terms of GDP generated.

In 2008, after significant community consultation, a new strategic plan called 'Sydney 2030' was developed, which set ambitious environmental targets of reducing emissions by 70 % by 2030. The city also adopted a renewable energy target of 30 % by 2030. Subsequently, a significant amount of work was undertaken to determine how the city would be able to reach this target, including the modelling of key emissions reduction technologies and interventions.

6.3.1.1 Eco-Claim

In 2008, the City of Sydney became the first local government to claim carbon neutrality.[11] It was based on a carbon audit undertaken by Hyder and was largely achieved through switching to GreenPower and purchasing carbon offsets.

[11] For more information, see http://www.sydneymedia.com.au/html/3689-australia39s-1st-carbon-neutral-government.asp?orig=Home.

Fig. 6.8 City of Sydney local government area. *Credit* © The Council of the City of Sydney (2011)

In November 2011, the City of Sydney became the first and only officially certified LGA under the Federal Government's National Carbon Offset Standard's (NCOS) Carbon Neutral Program.[12] By achieving this official status, the city's carbon neutrality claim is well substantiated and given greater credibility, as the audit and management follows a very rigorous process as required by the Federal Government Department of Environment, which administer the program. It is important to recognise that the status is only applied to the council's operational activities and not to the entire community, which the other case studies generally include.

The city's carbon footprint is similar to the other case studies in that it does not take into account the emissions associated with the embodied emissions in materials, although it does account for the emissions associated with subcontract work, which may involve construction activities.

---

[12] The media release is available here http://www.sydneymedia.com.au/html/4772-its-official-sydney-is-first-carbon-neutral-council.asp.

6.3.1.2 Low-Carbon Initiatives

The City of Sydney has many low-carbon initiatives in place and a well-developed 'Green Infrastructure Plan', which outlines the course of action over the coming years. Below is a list of some of the current and previous measures that have been implemented, followed by future planned measures.

Current and past initiatives

- Significant lighting retrofits throughout council-owned facilities (i.e., delamping, replacement of old lights with LEDs and other energy-efficient lights, lighting controls and motion sensors);
- Air conditioning and heating upgrades and better controls and management
- Installation of variable speed drives (VSD);
- Installation of solar photovoltaic panels on council-owned buildings.

Planned Initiatives

- Decentralised energy generation, with tri-generation systems planned at the precinct scale in numerous locations throughout the city;
- Solar photovoltaics to be installed at council sites;
- Wide-scale LED lighting retrofit, replacing 6450 street lights with energy-efficient LED lighting;
- Fleet eco-driver program and biofuels;
- Energy and water efficiency retrofit of the city's buildings and operations;
- Automated waste collection systems to be implemented, reducing transport emissions from garbage collection and increasing recycling efficiency;
- Advanced waste treatment (AWT) system to increase recycling.

6.3.1.3 Success

The first comprehensive carbon emissions audit was undertaken in 2006/2007, which has set the baseline for the emissions reduction measures. Several targets were then established including the ambitious 70 % emissions reduction target by 2030 based on 2006 levels (City of Sydney 2011).

When the city initially claimed carbon neutrality in 2008, the reduction was met largely through the purchase of GreenPower (100 % renewable electricity) and offsets. Nevertheless, the City of Sydney had also identified and has been implementing significant emissions reduction measures since the initial carbon audit in 2007. Fifty-three actions were recognised in the City's Environmental Management Plan (City of Sydney 2007), the majority of which have an impact on the city's emissions. The city met its 2012 target of a 20 % reduction on 2006 levels and is pursuing a range of further reduction activities (City of Sydney 2012).

In 2011, the city sought and became the first officially certified carbon-neutral government in Australia under the Federal Government's NCOS Carbon Neutral Program. The program requires continual demonstration of emissions reduction measures (as opposed to purely offsetting). In the light of this, as well as for financial reasons, the city decided to redirect the funding that was going towards GreenPower into installing city-owned renewable energy infrastructure within the city.

The City of Sydney currently remains the only government certified as Carbon Neutral under the NCOS standard.

6.3.2 City of Fremantle—Western Australia

Fremantle is a port city in Western Australia located on the mouth of the Swan River (see Fig. 6.9), 20 km from Perth, the state's capital city. It has a residential population of approximately 28,000, though as a local employment and tourist hub, its daily population is much higher.

The City of Fremantle has had a long history of being an environmental leader, having won numerous environmental awards over the years. In 2009, the City of Fremantle elected a new mayor based largely on his sustainability credentials and the city began adopting stronger environmental targets and implementing a greater range of sustainability measures. In 2009, the city also became the first carbon-neutral government in Western Australia and the second in Australia, following Sydney. In 2014, it became certified as a 'One Planet Council' under the BioRegional, One Planet Living Framework. This framework has 10 principles, one of which is zero carbon.[13]

6.3.2.1 Eco-Claim

Fremantle's carbon-neutral status and claim was originally based on, and compliant with, the ICLEI framework developed specifically for local governments. This framework focuses on the emissions generated from the local government's own operations (i.e. not those of the wider community). However, the council now uses the Federal Government's National Greenhouse and Energy Reporting Scheme (NGERS) Framework, which accounts for Scope 1 and Scope 2 emissions, though Scope 3 emissions are also included. While the status is not officially certified (i.e. under NCOS), the City of Fremantle's emissions inventory, or carbon footprint, has been verified by an independent third party, which brings a degree of credibility to the claim.

6.3.2.2 Low-Carbon Initiatives

The city has implemented a variety of emissions reduction measures over the years. Some have directly reduced the council's own operational emissions, and some

---

[13] For more information see http://oneplanetliving.org.au/one-planet-living/the-10-principles/.

Fig. 6.9 Aerial view—City of Fremantle. *Credit* LoveFreo (2013)

have targeted the wider community, but the latter are not included as part of the council's carbon footprint. Below are some of the more prominent initiatives that have been implemented over several years.

Current and Past Initiatives

- Developed a fully integrated carbon budget and carbon accounting system to measure and monitor operational carbon emissions;
- Implemented a shallow geothermal pool heating (300 kWe groundwater source heat pump and a shallow geothermal borefield combined with a 76 kWe cogeneration unit);
- Installing over 40 kW of solar arrays;
- Renewable Energy Investment Fund—1 % of city rates put aside for low-carbon initiatives and renewable energy;
- Purchasing a portion of energy for council operations from natural power (from 100 % renewable, emission-free energy);
- Upgrading city-owned public lighting to LED or sustainable technologies;
- Upgrading the city's computers with desktop LCD screens, small energy packs and other energy efficiency appliance initiatives;
- Increasing recycling from the council's administration office (more than doubling the original rate);
- Sending no organic material to landfill through a 'Waste Not' program;
- Sending Fremantle's community waste to the South Metropolitan Regional Council's (SMRC) Regional Resource Recovery Centre (RRRC), which diverts waste away from landfill, thereby reducing greenhouse gas emissions.
- Providing a free central area transport (CAT) bus service around Fremantle, encouraging the use of public transport;
- Decreasing the city's vehicle fleet by 10 cars and converting 18 vehicles to LPG;
- Creating a car share policy and facilitating the entry of a car share company (early 2015) to ensure that the city does not have to increase its vehicle pool to accommodate new staff members;

- Installing free charging points for electric vehicle in a council-owned car park;
- Quadrupling investment in bicycle infrastructure and providing bikes for council staff to attend local meetings;
- Helping to establish the award-winning Living Smart program, which is now run by community members and hosted at the City of Fremantle; and

Planned Initiatives

- Installing lighting timers for council buildings and facilities;
- Expanding a renewable and low-carbon energy program at the council-owned leisure centre (i.e. solar, geothermal and co/tri-generation projects and energy efficiency/monitoring projects);
- Developing a full energy strategy for council-owned facilities;
- Facilitating a range of community-oriented low-carbon initiatives.

6.3.2.3 Success

The city has been very successful in implementing environmental and emissions reduction initiatives over the years. This is most likely influenced from its active sustainability community, and a proactive council.

Initially, there was some controversy over the council's carbon-neutral claim, as it was not made clear that the carbon-neutral status only referred to the council's own operational emissions. There was also confusion over how the council achieved carbon neutrality, i.e. what emissions were calculated and what types of offsets were purchased. In response and after significant community consultation, the city released a Low Carbon City Plan (City of Fremantle 2011), which detailed the actions and measures the local government was taking in reducing emissions, and how it was claiming carbon neutrality. This Plan will be superseded by the City's One Planet Strategy due for release in 2015.

The City of Fremantle identified two specific targets to reach by 2020: firstly, a 20 % reduction in emissions through energy efficiency, and secondly, a 20 % reduction in emissions through switching to renewable sources of energy, all the while maintaining carbon-neutral status. The 40 % reduction equates to a projected 3000 tCO_2e/year reduction from now until 2020 (City of Fremantle 2011).[14]

6.3.3 North Port Quay—Western Australia

North Port Quay (NPQ) was a proposed coastal island development located near Fremantle, Western Australia, with a bold vision to be 'carbon free' and provide an international benchmark in sustainability. The proposed development was to be

[14] For more information see http://www.fremantle.wa.gov.au/sustainability.

Fig. 6.10 Aerial view of proposed position of NPQ. *Credit* Benchmark Projects Australasia

located on the seabed attached to the Fremantle harbour and will cover a total land area of 245.5 ha, protected by a 3.5-km breakwater (see Fig. 6.10), which will also provide protection from erosion to the current beaches. It was proposed to be a mixed-use development providing over 10,000 residential dwellings. The islands were to be completely powered by renewable energy and employ new and innovative low-carbon options for managing resources.

6.3.3.1 Eco-Claim

North Port Quay concept has had numerous carbon claims attributed to it, among them 'carbon neutral' and the more ambitious title 'carbon free'. The developers contended that it would be a self-sustaining port city and first carbon-free development in the world.[15] Though, as with most ambitious claims, scepticism soon arose with many questioning the reliability of the claim, particularly regarding the emissions associated with the creation of an island (Fernandes 2009). Nevertheless, the developers maintained that the measures taken to reduce emissions cover the cradle-to-grave life cycle of the island. A subsequent research report and analysis undertaken by Curtin University proved it could be done.

[15] See http://www.northportquay.com.au.

6.3.3.2 Low-Carbon Initiatives

In order to be carbon free, North Port Quay development intended to implement a variety of low-carbon initiatives. The initiatives would not only reduce the operational carbon footprint of the site, but will also help to offset other unavoidable emissions associated with development, making the entire development carbon neutral approximately 11 years after construction and carbon positive thereafter. These initiatives included the following:

- using low-carbon, local materials for the creation of the island including dredging spoil transported to the site using barges;
- energy and water efficiencies ranging across thermal design, appliances and lighting within buildings;
- 69 MW of solar PV on rooftops;
- 10.8 MW of commercial-scale wind;
- improved public transport (including a light rail connecting to Fremantle) and 100 % electric vehicle uptake by NPQ residents;
- waste to energy through incineration and biodigestion; and
- recycled water for irrigation, toilet and laundry uses.

6.3.3.3 Success

The concept for North Port Quay was proposed in 2008 by a consortium of land developers eager to showcase a state-of-the-art, sustainable city concept.

While the project's sustainability vision, ambition and intent were commendable, the project nonetheless attracted significant public and political opposition and became a highly controversial issue in Western Australia at the time. The main issues concerned the location of the development, with many vehemently opposing the creation of an island. There was concern over the environmental degradation of the seabed during construction, the unquantified amount of GHGs that would result from the creation of the island, the degradation of current beaches and reduced amenity for the current local community (Kerr 2011).

Many of the concerns were addressed during public consultation and were available under a 'FAQ' section on the project's Website.[16] Modelling and research undertaken within the ARC Linkage Project 'Decarbonising Cities and Regions' also found that NPQ could indeed become not only carbon neutral, but also carbon positive, providing enough carbon-free renewable electricity to power the island as well as feeding additional electricity into the grid to power 9300 homes in the neighbouring community (Newman et al. 2012). The development would be able to offset the emissions associated with the embodied energy in the materials used for

---

[16] See http://www.northportquay.com.au/FAQs/sample-faqs.html.

Fig. 6.11 Artistic image of Curtin University—Greater Curtin Master Plan. *Credit* Andy Sharp

the creation of the island and construction of buildings and infrastructure onsite in approximately 11 years.

Independent market research showed community support for the project sitting at approximately 70 % in favour of the project; however, due to significant political opposition,[17] the consortium decided to shelve the project indefinitely (Macdonald 2012).

Innovative low-carbon developments like NPQ are often difficult to achieve because they are so unusual and they immediately hit significant political and institutional barriers. Though, already in Perth, Western Australia, a range of new low-carbon precinct proposals could be seen to have built on the shoulders of NPQ. These developments include the proposal for a new innovative, low-carbon city on the campus of Curtin University (see Fig. 6.11).[18] This book aims to try and assist the next demonstrations to be easier through mainstreaming mechanisms.

6.4 Comparing Developments

6.4.1 Variations in Claims

As evidenced from the case studies above, the way in which each development, local council or city describes and asserts its carbon claim varies greatly. While some have a broader focus on sustainability (e.g., Vauban and B001), others have more carbon-

[17] There are differing opinions about the level of support shown by the community.

[18] For more information about Greater Curtin, visit http://properties.curtin.edu.au/whoweare/masterplan.cfm.

specific aspirations and targets (e.g., Masdar, NPQ, City of Sydney and Fremantle). For the developments that have a greater focus on carbon, further variations exist around the approach taken to measure emissions, i.e. whether they have adopted an operational perspective or a life cycle approach, and thus the sources of emissions they include in their analysis. Each development reported on various carbon-related aspects, but the areas seemed to be chosen based on what was relevant to each individual case (i.e., as a way to demonstrate something they did well).

6.4.2 Emissions Sources Included in Claims

In order to illustrate the differences between the emissions sources included in the case studies discussed, each case has been compared on the basis of six areas of emission sources. The results are demonstrated in Table 6.1. The analysis is based purely on whether measures were implemented within the case studies that addressed the selected areas, not on whether they actually calculated the emissions associated with that area.

Based on this comparison, energy and transport are the two most targeted areas. Every development targeted stationary energy in one form or another.

Transport was also mentioned relatively consistently, with many developments prioritising access to public transport, as well as ensuring that sufficient amenity existed close by and thus helping to discourage unnecessary travel by residents. Nevertheless, this is not consistent among all, nor was it clear how transport emissions were calculated. Some, like Dongtan, may indeed be much worse on transport than business-as-usual development.

Table 6.1 Comparison of emission sources within case studies

	Embodied Emissions	Construction	Energy	Water	Waste	Transport
BedZED	✔	✔	✔	✔	✗	✔
Vauban	✗	✗	✔	✗	✗	✔
B001	✗	✗	✔	✔	✔	✔
Hammerby Sjostad	✗	✗	✔	✔	✔	✔
Masdar	✔	✔	✔	✔	✔	✔
Dongtan	✗	✔	✔	✗	✔	✔
City of Sydney	✗	✗	✔	✓	✔	✔
City of Fremantle	✗	✗	✔	✗	✔	✔
North Port Quay	✔	✔	✔	✔	✔	✔

Waste is also seen as relatively important. The two areas that were least considered were the construction process and embodied emissions in the materials, despite the growing percentage of emissions attributable to this area because of the improvements in operational energy efficiency (Sturgis and Roberts 2010). This demonstrates that few developers or local councils adopt a full life cycle approach when targeting or reporting emissions.

Water was also frequently targeted, but the focus was often due more to the ecosystem benefits than emissions reduction opportunities.

The terminology used to describe the claims is also inconsistent, and common terms are not always well defined. Without having a set of clearly communicated and universally accepted definitions, terms will continue to confuse and lack meaning. This, together with the absence a common metric or framework for undertaking the carbon analysis, has made comparisons between the developments difficult and is likely to continue to result in suspicion and distrust in carbon claims, as evidenced in the cases of NPQ and Dongtan. Furthermore, the lack of follow-up and ongoing evaluation also adds a layer of controversy and complexity as claims can quickly become outdated and inaccurate, as shown in the case of BedZED. In Australia, false or misleading claims are liable for fines of up to $1.1 million from the Australian Competition and Consumer Commission (Australia Competition and Consumer Commission [ACCC] 2008).

6.4.3 Inconsistent Reporting

Another issue, which can lead to inaccurate claims, relates to the way case studies are documented. Ambitious, progressive and unprecedented case studies tend to draw significant attention and are widely documented and described in the literature. However, as stories are reiterated and retold, facts, statistics and descriptions often become distorted with little follow-up or investigation into the claims being carried out. This was evident within this literature review, where conflicting numbers relating to emissions reductions were presented without original data sources being provided (e.g., in the case of Takahashi and Onishi 2011). This can lead to varying, confusing data and exaggerated claims.

Williams (2009) acknowledged this problem and, using Dongtan as an example based on the considerable attention it drew worldwide, suggested that figures are often picked up and used without sufficient background research or investigation of where the figures came from, how goals are intended to be achieved or whether they are indeed being achieved. Dongtan thus provides an apt example of the need for tangible carbon metrics that can demonstrate how even a proposed development is going to meet its carbon goals.

6.4.4 Problems with Scaling up Low-Carbon Developments

From the case studies examined, it appears there may be some additional issues associated with scaling up low-carbon developments (Nader 2009). Most of the developments examined had a population less than 10,000. Masdar and Dongtan are examples of attempts to create larger-scale eco-cities, but both have experienced difficulties. Masdar appeared to run out of funds after having completed only the first stage of the development, while Dongtan, which was going to be one of the largest eco-developments built, never actually began construction.

With increasing scale, comes increasing complexity, particularly around the management of resources. Currently, larger developments generally consume more than they can generate onsite. While this may not necessarily be the case into the future, we are yet to be presented with an example of an ideal, broadly replicable large-scale, low-carbon development. However, Masdar, although often considered an unrealistic example of a low-carbon city due to the large financial costs associated with the development, is nonetheless likely to be a great test bed for new low-carbon technologies that can cater for larger populations, particularly if the development reaches the second stage. The fine-grained data currently being collected by the ongoing monitoring and evaluation process of the largely experimental city will help to inform and develop improved future designs of low-carbon cities.

6.4.5 Direction Needed

Regardless of the size, having frameworks and processes in place to direct developers to the areas that need to be targeted in terms of overall design and management of resources can help to create greater understanding of what makes a truly low-carbon development.

6.5 Conclusion

From the case studies presented in this chapter, it is clear that significant variation currently exists in what developers target in terms of emissions reduction and, thus, which sources of emissions have been included in achieving their low-carbon status. The lack of a common metric or framework for conducting carbon analyses, together with inconsistent terminology and lack of universally accepted definitions, makes it difficult to compare developments and their claims. This can lead to suspicion and distrust in carbon claims, as evidenced in the North Port Quay example. Furthermore, without continuous follow-up and evaluation, claims can quickly become outdated and inaccurate.

This chapter highlights the need for a universal carbon accounting framework for precinct-scale development that identifies the specific areas that need to be considered and targeted by developers making carbon claims, as well as a standardised approach and methodology for quantifying those emissions. This will help to make carbon claims and assertions much more meaningful and comparable, bring greater credibility to the concept of low-carbon developments and thereby increase the opportunity for them to become mainstreamed. Without this, certification will never be possible.

The next chapters examine how urban development can be certified and/or recognised for its carbon reduction, and what mechanisms can assist in mainstreaming the concept.

References

Australia Competition and Consumer Commission [ACCC]. (2008). *Carbon claims and the trade practices act*. Retrieved from http://www.greenwashreport.org/downloads/ACCC_Carbon.pdf.

City of Fremantle. (2011). *Low carbon city plan: 2011–2015*. Retrieved from http://www.fremantle.wa.gov.au/files/1f8ee5d0-d2e9-4712-acfd-9f9800ea66f6/Low_Carbon_City_Plan_2.pdf.

City of Sydney. (2007). *Environmental management plan*. Retrieved from http://pandora.nla.gov.au/pan/92424/20081217-1140/www.cityofsydney.nsw.gov.au/Environment/documents/CityofSydneyEMP.pdf.

City of Sydney. (2011). *Sustainable Sydney 2030: Community strategic plan (2011)*. Retrieved from http://cdn.sydney2030.com.au/documents/Community-Strategic-Plan-2011.pdf.

City of Sydney. (2012). *NCOS carbon neutral program—Public disclosure summary. Sydney*. Retrieved from http://www.cityofsydney.nsw.gov.au/__data/assets/pdf_file/0018/131913/2010-11NCOSPublicDisclosureSummary.pdf.

Dac & Cities. (2012). *Dongtan: The World's first large-scale eco-city?* Retrieved 04/10/12, from http://www.dac.dk/en/dac-cities/sustainable-cities-2/show-theme/energy/dongtan-the-worlds-first-large-scale-eco-city/?bbredirect=true.

Engwicht, D. (1992). *Towards an Eeo-City: Calming the Traffic*. Sydney, Australia: Envirobook.

Fernandes, A. (2009). North Port Quay environmental claims misleading, says expert. Retrieved from http://www.sciencewa.net.au/topics/environment-a-conservation/item/854-north-port-quay-environmental-claims-misleading-says-expert.html.

Hodge, J., & Haltrecht, J. (2009). *BedZED Seven years on: The impact of the UK's best known Eco-village and its residents*. BioRegional.

Joss, S., Tomozeiu, D., & Cowley, R. (2011). Eco-Cities—A global survey 2011: Eco-city profiles. *International Eco-Cities Initiative*. London: University of Westminster.

Kansara, T., & Ridley, I. (2012). Post occupancy evaluation of buildings in a zero carbon city. *Sustainable Cities and Society, 5*, 23–25.

Kerr, T. (2011). *Negotiating green urbanism in imagined communities*. Retrieved from http://www.inter-disciplinary.net/wp-content/uploads/2011/06/kerrepaper.pdf.

LoveFreo. (2013). *Fremantle by Helicopter*, from http://www.lovefreo.com/2013/04/29/fremantle-by-helicopter/#!prettyPhoto.

Macdonald, K. (2012). *Ambitious Marina still on the cards* (p. 14). Perth: The West Australian.

Masdar City. (2011). Sustainability in Masdar City. Retrieved from http://www.masdar.ae/en/masdar-city/detail/sustainability-in-masdar-city.

Medeazza, G., & Moreau, V. (2007). Modelling of water-energy systems: The case of desalination. *Energy, 32*, 1024–1031.

Monbiot, G. (2006). Smoke and mirrors. *New Scientist, 192*(2583–2584), 25.

Nader, S. (2009). Paths to a low-carbon economy—the Masdar example. *Energy Procedia, 1*, 3951–3958.

Newman, P., & Jennings, I. (2008). *Cities as sustainable ecosystems: Principles and practices*. Washington, DC: Island Press.

Newman, P., Beatley, T., & Boyer, H. (2009). *Resilient cities: Responding to peak oil and climate change*. Washington, DC: Island Press.

Newman, P., Rauland, V., Bunning, J., & Beattie, C. (2012). *North Port Quay: A world class carbon neutral development*. Perth: Industry Report. CUSP.

Nilsson, A., & Elmroth, A. (2005). The buildings consume more energy than expected. In *Sustainable city of tomorrow: Experiences of a Swedish House exposition*. (pp. 108–109). Stockholm: Swedish Research Council for Environment, Agricultural Sciences and Spatial Planning.

Reiche, D. (2009). Renewable energy policies in the Gulf countries: A case study of the carbon-neutral "Masdar City" in Abu Dhabi. *Energy Policy, 38*(1), 378–382.

Roseland, M. (1997). *Eco-City dimensions: Healthy communities, healthy planet*. Gabriola Island, Canada: New Society Publishers.

Snow, D. A., Soule, S. A., & Kriesi, H. (2004). *The Blackwell companion to social movements*. New York: Wiley.

Sturgis, S., & Roberts, G. (2010). *Redefining zero: Carbon profiling as a solution to whole life carbon emissions measurement in buildings*. London: RICS Research.

Takahashi, K., & Onishi, T. (2011). Moves in Europe toward the creation of low-carbon cities. In T. Onishi & H. Kobayashi (Eds.), *Low carbon Cities: The future of urban planning*. Kyoto: Gakugei Shuppan-Sha Co., Ltd.

The Oresund Committee. (2009). *The Öresund Region—A growth centre for environmental technology*. Retrieved from http://www.kooperation-international.de/fileadmin/public/cluster/Kopenhagen/klimatbroschyrENweb.pdf.

Williams, A. (2009). *Dongtan: the eco-city that never was Spiked*. Retrieved from http://www.spiked-online.com/newsite/article/7330.

Williams, J. (2012). *Zero carbon homes: A road map*. New York: Earthscan.

World Bank. (2009). *Climate resilient cities: A primer on reducing vulnerabilities to disasters City Profiles: Dongtan, China*. Retrived from http://siteresources.worldbank.org/INTEAPREGTOPURBDEV/Resources/573631-1233613121646/dongtan_extop.pdf.

World Commission on Environment and Development [WCED]. (1987). *Our common future*. New York: Oxford University Press.

Wu, F. (2012). China's eco-cities. *Geoforum, 43*, 169–171.

Chapter 7
The Rise of Carbon Neutrality

7.1 Introduction

The concept of carbon neutrality emerged at a time of exceptionally high public awareness of climate change due to numerous government reports, e.g. the IPCC Fourth Assessment Report in 2007, the UK Stern Review in 2006 and Australia's Garnaut Review in 2008, and prominent public figures such as Al Gore (An Inconvenient Truth in 2006). People and businesses eager to contribute and take action on climate change readily embraced the idea of carbon neutrality, which involved decreasing emissions as much as possible internally, before purchasing offsets to reduce the remaining emissions to zero.

Countless companies and organisations around the world soon began offering the option of becoming carbon neutral, either for their businesses or for their personal emissions. Online carbon calculators started to proliferate on the internet, allowing people to calculate their carbon footprint, offset their emissions and become carbon neutral all in a matter of minutes (Murray and Dey 2008). Businesses and organisations soon offered carbon-neutral certification, providing a much sought-after branding opportunity for many carbon-neutral customers. The concept also became popular among local governments, cities and land developers keen to promote their green credentials.

Recognising the need and increased demand for credible offsets, the UK government set up the Carbon Trust in 2001 and the Australian Federal Government introduced the 'Greenhouse Friendly' initiative, which ran from 2001 until 2012.[1] Greenhouse Friendly allowed businesses to market carbon-neutral products and reward greenhouse gas abatement within various sectors of the economy (DCCEE 2011a).

Carbon neutrality is now available for products, services and activities such as car use, air travel, household energy consumption and entire business activities

[1] Greenhouse Friendly has since been replaced with the National Carbon Offset Carbon Neutral Program, which commenced in July 2010 and is discussed further in Chap. 11.

V. Rauland and P. Newman, *Decarbonising Cities*, Green Energy and Technology, DOI 10.1007/978-3-319-15506-7_7

(Wiedmann and Minx 2008). Carbon neutral has even begun to permeate public policy, featuring in the 2007 UK Climate Change Bill, which required all new residential buildings from 2016 to be carbon neutral[2] (BBC News 2006; UK Department for Communities and Local Government 2007). The concept became so popular that in 2006, 'carbon neutral' became the New Oxford American Dictionary's Word of the Year.[3]

This chapter examines what carbon neutrality is and addresses climate change. It goes on to highlight some of the main issues that have emerged as a consequence of its rapid growth. The main issue has been the lack of guidance around carbon accounting, particularly for those working within the built environment and specifically at the precinct scale. From the literature, a list of key considerations is provided, which, it is argued, should be taken into account and publicly disclosed by anyone making a carbon claim.

7.2 Defining Carbon Neutrality

Despite the widespread use and prevalence of the term carbon neutrality within the public domain, there remains no universal definition, nor any widely accepted international certification system. As Murray and Dey (2008) note, the term currently appears to be 'defined by popular usage' (238). Table 7.1 lists nine definitions of carbon neutrality.

As evidenced by the definitions above, the broad intention of carbon neutrality is to create a balance of carbon in the atmosphere in relation to the inputs and outputs of a product or service, or in the context of this book, a community, precinct or a city.

It generally involves reducing emissions as much as possible through means such as energy efficiency measures, or in relation to the built environment, it may be through better building design, transport infrastructure, behaviour change or switching to renewable sources of energy, before offsetting the remaining, unavoidable emissions (e.g. emissions embodied in materials of buildings and infrastructure) (DCCEE 2011b; Newman et al. 2009).

The process of achieving carbon neutrality consists of three main steps as illustrated in Fig. 7.1: measuring, reducing and offsetting emissions (DCCEE 2011b; Department of Communities and Local Government 2009).

Calculating emissions is the first fundamental step in achieving carbon neutrality or any carbon reduction goal. As is commonly stated by those working in the field, 'you can't manage what you don't measure'. Identifying emission sources and

[2] This term later changed to become 'zero carbon'. The requirement has also been eased to a more voluntary standard.

[3] See: http://www.climatesciencewatch.org/2007/01/08/carbon-neutralnew-oxford-american-dictionarys-2006-word-of-the-year/.

Table 7.1 Selected definitions of carbon neutrality

Source	Definition of carbon neutrality	Reference
Oxford Dictionary	Making or resulting in no net release of carbon dioxide into the atmosphere, especially as a result of carbon offsetting	Oxford Dictionary. (no date). Definition of Carbon Neutral. Retrieved from http://oxforddictionaries.com/definition/english/carbon-neutral?q=carbon+neutral [Accessed 15/03/12]
Wikipedia	Achieving net zero carbon emissions by balancing a measured amount of carbon released with an equivalent amount sequestered or offset, or buying enough carbon credits to make up the difference	Wikipedia. (no date). Carbon Neutral. Retrieved from http://en.wikipedia.org/wiki/Carbon_neutral [Accessed 15/03/12]
Low Carbon Australia	A situation where the net emissions associated with a product or an organisation's activities are equal to zero through the reduction of emissions and the acquisition and cancellation of carbon offsets that meet stringent criteria, to offset the remaining emissions	Low Carbon Australia. (2011). Carbon Neutral Program Guidelines. Retrieved from http://www.climatechange.gov.au/government/initiatives/low-carbon-australia/ncos-carbon-neutral-program/carbon-neutral-program-guidelines.aspx [Accessed 22/08/12]
ICLEI	A state where no net emissions are produced by a particular entity or activity during a particular time period	ICLEI. (2008). Carbon Neutrality Framework for Local Governments. Retrieved from http://www.fremantle.wa.gov.au/files/223c5cc1-0c26-4b73-a4a6-9dac00ee6cad/A_ICLEI_Carbon_Neutrality_Framework_081031.pdf [Accessed 09/01/15]
UK Department of Energy and Climate Change	Carbon neutral means that—through a transparent process of calculating emissions, reducing those emissions and offsetting residual emissions—net carbon emissions equal zero	Department of Communities and Local Government. (2009). Guidance on Carbon Neutrality. Retrieved from http://www.decc.gov.uk/publications/DirectoryListing.aspx?tags=3
Murray and Dey (2008)	Cancelling out the harm done to the earth's atmosphere by one type of greenhouse-gas-generating human activity, through another human activity that either reduces CO_2 emissions by an equal amount or prevents an equal amount being generated by an essential CO_2 producing activity by substituting a non- or low-carbon producing alternative	Murray, J., and Dey, C. (2008). The carbon neutral free for all. *International Journal of Greenhouse Gas Control* 3:237–248

(continued)

Table 7.1 (continued)

Source	Definition of carbon neutrality	Reference
Rauch and Newman	…no net carbon emitted to the atmosphere from institutional operation due to zero net fossil carbon use	Rauch, J, N., and Newman, J. (2007), Defining sustainability metrics in an institutional setting. Retrieved from http://sustainability.yale.edu/sites/default/files/rauch_newman_2009.pdf [Accessed 26/10/12]
Zuo et al. (2012)	Carbon emissions from the energy consumption for materials and construction process, operational energy use, and those from tenant activities associated with building such as transportation requirements	Zuo, J., Read, B., Pullen, S., and Shi, Q. (2012). Achieving carbon neutrality in commercial building developments—Perceptions of the construction industry, *Habitat International*, 36: 278–286
Carbon Neutral Company	Carbon neutrality, or having a net zero carbon footprint, refers to achieving net zero-carbon emissions by balancing a measured amount of carbon released with an equivalent amount sequestered, avoided or offset	Carbon Neutral (UK Private Offset company). Retrieved from http://www.carbonneutral.com/knowledge-centre/carbon-glossary/

calculating an initial carbon footprint is the first critical step in managing emissions and is essential to the understanding and recording of where reductions can be made. The next step involves implementing measures to reduce emissions to the extent possible, before finally offsetting residual, unavoidable emissions. The process is generally undertaken on a regular (usually annual) basis in order to calculate and document the reductions.

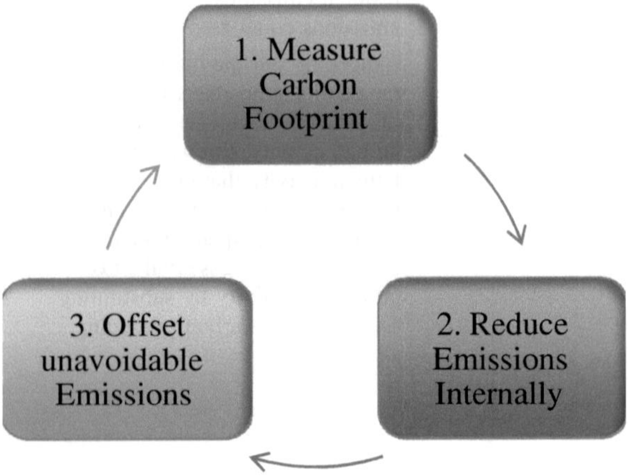

Fig. 7.1 Graphical representation of carbon neutrality

Despite apparent consensus over the broad definition of carbon neutrality and the process involved, there remains significant variation in how the task is approached specifically, for example, how the carbon footprint is calculated (e.g. what sources are included) and to what extent emissions are reduced before being offset. This is discussed further in Sect. 7.4.

7.3 Benefits of Carbon Neutrality

While the motivation for pursuing carbon neutrality may have originally stemmed from environmental concerns and the desire to promote green credentials, there are several other reasons why businesses and individuals are now choosing to go carbon neutral.

To begin with, understanding and reducing your carbon footprint—the first two steps in achieving carbon neutrality—can help save considerable costs on utility bills by identifying areas where wastage is occurring (see Box 1). Many individuals and businesses are often unaware of the inefficiencies associated with their homes, office buildings and operational activities. When the National Greenhouse and Energy Reporting Scheme (NGERS) began in Australia in 2008 (a new mandatory reporting scheme for large businesses), this was the first time many businesses had undertaken energy audits. The first reporting from consultants hired to do these audits found remarkably easy savings, e.g. one office building in Perth found they had been paying the power bill for their neighbours for the past 20 years.

Undertaking energy and carbon audits helps to understand where the biggest sources of emissions are, which in turn helps to identify potential problems areas and where to prioritise reduction efforts. Identifying leaks and inefficient appliances and equipment, as well as choosing the right electricity tariff, can save substantial costs on utility bills. Investa Property Group, a leading Australian real estate company, demonstrated how they managed to cut over $5 million in utility costs from better managing their buildings and facilities (Roussac and Kuiper 2009). For many businesses, these financial savings can be significant and are often the primary motivation for seeking carbon neutrality.

Box 1. South Fremantle Senior High School: The First Certified Carbon Neutral School in Australia

In 2012, South Fremantle Senior High School (SFSHS), a public school located in Western Australia, celebrated a momentous achievement: they became the first certified Carbon Neutral School in Australia under the Australian Federal Government's NCOS Carbon Neutral Program.

The school had been working towards this goal since 2007 when Curtin University Sustainability Policy (CUSP) Institute's Director, Professor Peter Newman, and father of a student at the time, proposed the idea. The idea was readily embraced by the principal and several key community members and soon after created a Carbon Neutral Working Group (see Fig. 7.2), which has met fortnightly since its inception and has been paramount to the project's success.

A Carbon Neutral Project Officer was employed to drive the carbon reduction initiatives including energy efficiency retrofits and renewable energy measures, all of which led the school to reduce emissions by approximately 15 % from 2007 levels. These savings were enough to cover the salary of the part-time Carbon Neutral Project Officer. In 2008, the school switched their electricity to GreenPower—100 % renewable energy—eliminating the emissions arising from traditional coal-fired power generation, dramatically reducing their overall emissions. This was supplemented with the planting of around 30,000 trees in the wheat belt by students each year.

In 2011, the school decided to seek official carbon-neutral certification. Professor Peter Newman, who remains a key player within the committee, approached PhD candidates at the time, Samantha Hall and Vanessa Rauland, to see if they would assist with the certification process based on their knowledge of carbon accounting and GHG reporting. Together, they undertook a comprehensive carbon audit of the school, ensured the school purchased certified carbon offsets to eliminate the remaining emissions and helped them through the certification process.

In May 2012, the school successfully gained the official carbon-neutral certification through Low Carbon Australia, making it the first school and not-for-profit organisation in Australia to be certified. The achievement was celebrated with the Prime Minister in September 2012. At the event, the Prime Minister warmly expressed her appreciation that a community-based initiative had now demonstrated how to save both greenhouse gas emissions and money. Vanessa and Samantha were both recognised for their contribution to the school's achievement (Fig. 7.3).

Going carbon neutral is a vehicle by which the school is fulfilling their vision. SFSHS's journey, including the establishment of a committee, the appointment of a Carbon Neutral Project Officer, the way they financed their program, how they are empowering staff and integrating carbon neutrality into all aspects of the curriculum, and thus providing innovative and practical learning opportunities for students, provides a valuable and inspiring model which can be applied to other schools.

Vanessa and Samantha are now exploring ways to help to make this a reality for all schools in Australia through a national research project on low-carbon, high-performance schools' within the CRC in Low Carbon Living.

There are also numerous co-benefits associated with reducing emissions, particularly within office environments. For example, emissions reduction measures, such as improving access to natural light and fresh air, not only help to reduce the

Fig. 7.2 SFSHS Carbon Neutral Working Group (includes the authors). *Credit* Kathy Anketell

Fig. 7.3 Dr. Vanessa Rauland receiving an award from the Prime Minister Julia Gillard. *Credit* Vanessa Rauland

electricity demand for artificial lighting and ventilation, but can also dramatically improve indoor environment quality (IEQ) (Fisk and Sepanen 2007; Hall 2014). Several studies and reports have linked improved IEQ to an increase in staff productivity (Fisk and Sepanen 2007; Kubba 2012), and some of the observed IEQ effects include fewer sick days, greater staff retention rates and improved employee productivity. Greener buildings also often benefit from increased market value and higher rents (Eichholtz et al. 2008).

Hall (2014) was able to show in several buildings in Perth that significant gains could be made when all stakeholders in the buildings were invited to participate in the regeneration of their buildings from an energy perspective. Health and productivity benefits considerably outweighed the energy cost savings.

Finally, the process of going carbon neutral can help to minimise vulnerability to the financial risks associated with carbon, water and electricity price increases, and future legislation. Managing and reducing carbon within many sectors of the economy will become more and more critical as the goal to reduce carbon by 80 per cent by 2050 becomes increasingly accepted and mainstreamed. For those not currently mandated to report and/or pay a carbon price, voluntarily choosing to go carbon neutral, and thereby calculating and reducing carbon emissions, will help prepare businesses for future regulatory and legislative requirements. These may include increasingly stringent building codes and the expansion of both mandatory energy disclosures for buildings. For many companies, the multiple benefits associated with pursuing carbon neutrality as outlined above simply make good business sense.

7.4 Carbon Neutrality: Carbonwash or the Real Deal?

Carbon neutrality is now applied globally to a range of emissions stemming from a variety of different activities. Nevertheless, despite the increased usage of the term and perhaps because of the wide variety of applications, there remains no universal definition, nor any widely accepted framework for calculating emissions or certifying achievements. Adding to the problem has been the emergence of a whole set of new terminology, as discussed in Chap. 6, such as carbon zero (or zero carbon), zero emission, carbon-free, low carbon, carbon negative, carbon positive, climate positive and climate neutral. Most likely the result of a desire to differentiate oneself in an increasingly ambitious world, the myriad of names and their varying definitions, or lack thereof, has unfortunately led not only to confusion, but also to scepticism and mistrust commonly represented by the term 'greenwash', or perhaps more appropriately, 'carbonwash'. The lack of consistency in definition and in the methodology for calculating emissions or how the carbon reduction is ultimately achieved (for example, whether offsets are used) remains an issue.

Nevertheless, new guidelines and protocols have gradually been developed to bring standardisation to carbon claims, as well as new carbon certification schemes and regulations around the development of offsets. However, these processes are not always applicable to all sectors wishing to pursue carbon neutrality. The built environment and urban development, which is the focus of this book, is one area that is not well defined. This sector lacks meaningful frameworks and systems for determining emissions, as well as emissions reductions, or ways of recognising achievements. Indeed, no successful model or program exists or is being developed that provides a standardised way to account for the carbon emissions, or a method for achieving carbon neutrality, specifically within the built environment, and in particular, for urban precinct-scale development.

7.5 Increasing the Credibility of Carbon Claims

In 2010, the Australian Federal Government introduced a new national voluntary scheme for certifying carbon neutrality (NCOS Carbon Neutral Program) in order to bring greater credibility to carbon-neutral claims (DCCEE 2011b).[4] They defined carbon neutrality as:

> a situation where the net emissions associated with a product or an organisation's activities are equal to zero through the reduction of emissions and the acquisition and cancellation of carbon offsets that meet stringent criteria, to offset the remaining emissions. (DCCEE 2011b: 1)

However, the guidelines remain broad and are more directed at organisations and services and thus fail to achieve any further clarification regarding emissions associated with many built environment activities such as land development. Nor do they take into account various other issues specific to this sector such as ongoing maintenance of carbon neutrality (who is ultimately responsible) and what offsets should be considered eligible.

In relation to offsets, significant changes occurred within Australia's political landscape over the period 2009–2014, such as its ratification of the Kyoto Protocol; the cessation of the Greenhouse Friendly Initiative and the introduction of the National Carbon Offset Standard (NCOS); the introduction of mandatory carbon pricing legislation (the Carbon Price Mechanism—CPM) along with a suite of other policies and measures including the establishment of the Carbon Farming Initiative (CFI)[5]; the subsequent repeal of the carbon price; and the introduction of the Direct Action Plan. These have all had important implications in relation to the use and eligibility of offsets used for carbon neutrality.

For example, when the CPM was in place, which targeted large energy users and heavy polluting industries, land developers seeking carbon neutrality under the federal government's scheme were unable to use their onsite emissions reductions (i.e. energy efficiency initiatives or renewable energy production) to generate offsets, which could counter emissions from other aspects of their development; this would have been seen as double counting. In the presence of a mandatory carbon price, the only eligible domestic offsets were those generated from activities under the CFI,

---

[4] In March 2013, the Department of Climate Change and Energy Efficiency merged with another department, forming the Department of Industry, Innovation, Climate Change, Science, Research and Tertiary Education (DIICCSRTE). With the change of government less than six months later, this got disbanded and the separate departments got split into the Department of Industry and the Department of Environment.

[5] The CFI essentially provided an opportunity for sectors not covered by the carbon price (i.e. agriculture) to participate in the carbon market by providing carbon credits or offsets for voluntary emission reduction activities. Credits were only awarded to projects from sectors not covered by the Carbon Price Mechanism. This has since change with the repeal of the carbon price; many more projects are now eligible under the CFI.

which were from sectors not liable under the CPM (i.e. agriculture). Thus, despite a political and popular desire to decarbonise urban development, this issue with offsets meant there was no coherent and useful certification process for achieving significant carbon reduction within the built environment.

The removal of the CPM in 2014 essentially removed the issue associated with double counting. Since the introduction of Australia's Coalition Government Direct Action Plan, a range of new projects, which were previously excluded under the CPM, are now permitted under the CFI. It is expected that these offset credits (which will generate ACCUs) will be able to be used to offset other emissions associated with an urban development under the NCOS Carbon Neutral Program.

In most nations where front-end carbon pricing and certification has been implemented, a similar problem exists: urban development demonstrating innovative carbon reductions will struggle to generate carbon credits, as it is often seen as double counting. However, there are potential ways to deal with this, and new end-user programs such as Australia's Direct Action Plan demonstrate how this market is possible. There are examples where the two schemes—a front-end and end-user carbon trading scheme—can exist alongside each other. This will be discussed later in the book.

7.6 Carbon-neutral Cities and Urban Development

Several cities and developments around the world have claimed carbon neutrality,[6] as seen in Chap. 6. Even regions and entire nations are claiming the status. In 2007, the Vatican pledged to become the first carbon neutral state (Lean and Kay 2008; Rosenthal 2007)[7] and it was joined by New Zealand, Costa Rica, Norway and Iceland, all pledging and competing to be the first country to become carbon neutral (Lean and Kay 2008). However, the way in which each has achieved, or intends to achieve their claimed status often differs considerably, if it is explained at all.

New terminology has evolved as people have sought to create a point of difference and explain their carbon achievement in a unique way. The lack of consistency in definition remains a problem, even among those working in the building industry (Zuo et al. 2012), as does the lack of clarity regarding the emission inventory process and how carbon neutrality (or the respective term) is ultimately achieved.

While considerable discussion is occurring around the carbon definitions related to buildings, there has been little coherent argument on what this might mean or

[6] This is often as an aspirational target or goal to be reached in the future.

[7] However, carbon neutrality was never realised, highlighting the seemingly ubiquitous issue of greenwashing (Struck 2010).

look like at the precinct or city scale—despite the growing industry focus on carbon reductions at this level (Ewing et al. 2008; Roseland 2012; The Climate Group 2010).

7.6.1 Precinct-Scale Carbon Claims

Countless developments around the world are demonstrating how to reduce carbon at the precinct level as the basis for demonstrating twenty-first-century urban innovation (Lehmann 2015). The importance of going beyond the individual building and considering precinct-scale urban land development is that it begins to consider the urban and regional structure and form in terms of infrastructure provision, taking into account both traditional centralised and distributed forms of resource management, as well as opportunities for greater public transport, walking and cycling infrastructure.

However, as Chap. 6 demonstrated, there is currently no consistency in the way to go about the process, how to measure and report emissions, or how to assert a carbon reduction or claim.

Most developments appear to concentrate predominantly on the carbon emissions associated with onsite energy production and use. Although many also take into consideration additional factors such as water and waste, these are not always represented in terms of carbon. The embodied emissions associated with the materials used in developments are rarely accounted for in carbon analyses, despite this being a growing area of emissions (Sturgis and Roberts 2010).

Transport appears to play a significant role in lowering the per capita carbon footprint of residents living within the case studies discussed, particularly when compared to the citywide average. However, again this is not consistently documented in all developments, nor is it clear how emissions have been calculated.

Boake (2008) suggests a three-step process to achieve carbon neutrality as demonstrated in Fig. 7.4.

The inconsistent terminology, together with the lack of a common metric or framework for the carbon analyses, or a common process to reduce emissions, makes it difficult to compare developments and cities in terms of their carbon reduction. Therefore, in order to make any such assertions meaningful, comparable and legitimate, a standardised approach to quantifying the carbon emissions arising from precinct-scale development needs to be developed and promoted, as well as an ongoing evaluation process.

The following sections look at some of the main issues currently associated with carbon-neutral claims and provide an analysis of current and proposed GHG accounting frameworks, methodologies and initiatives aimed at bringing standardisation to the quantification of emissions at the precinct and city scales. An examination of some of the issues surrounding offsets is also provided.

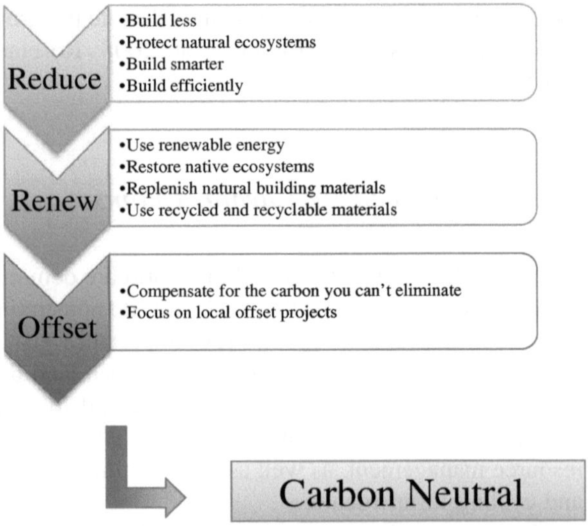

Fig. 7.4 Process to achieve carbon neutrality. Adopted from Boake (2008)

7.7 The Need for a Sector-Specific Framework

To date, there has been very limited research into carbon-neutral concepts and initiatives at the urban scale. The authors note the following as key issues for progressing the concept of carbon-neutral or low-carbon urban development:

1. There is currently no framework or guidelines, which sets out what emissions should be included in an analysis of low-carbon or carbon-neutral urban development. Countless rating tools and schemes exist that attempt to measure various aspects of sustainable development at the precinct level (AILA 2010), but most are limited to check lists and none focus solely on GHG emissions or offer a comprehensive or specific enough carbon accounting framework to recognise carbon neutrality;
2. Although several carbon-neutral certification schemes exist around the world, there has never been anything like the complexity of an urban subdivision, new precinct development or community-scale redevelopment submitted for certification. As a result, no carbon accounting framework or consistent process for certifying land development currently exists, meaning standards may vary between developments; and
3. There is no evaluation of such initiatives that can inform the planning process or enable the developer to gain from the experience. Thus, valuable information that could provide specialised guidance for future low-carbon and carbon-neutral development may be lost.

7.7.1 Accounting for Carbon in Urban Development

Despite the largely accepted, broad definition of the term carbon neutrality, the precise approach and methods required to undertake each specific step of the process, particularly in the case of urban development, remain unclear on several levels including:

- the specific GHG emissions that are covered (i.e. the types of gases);
- the various scopes (i.e. direct or indirect emission sources—also known as Scope 1, 2 and 3 emissions)[8] and boundaries of emissions; and
- whether or not carbon offsets are included and, if so, the degree to which emissions have been offset before being reduced.

Further detailed information around offsets is required, for example, whether they are produced onsite or purchased from a third party, and in the latter case, how credible the offset provider is (Department of Communities and Local Government 2009; Wiedmann and Minx 2008). These and other factors are outlined in Table 7.2.

The case studies analysed in Chap. 6 illustrate that developments currently do not address these questions systematically, resulting in significant variation in their carbon definitions, goals and achievements. A universal process for determining what factors need to be addressed and clarified when making carbon claims at the precinct level would therefore go a long way in helping to overcome these issues. Kennedy and Sgouridis (2011) also identify the lack of definition and need of a carbon accounting framework for urban development, stating:

> Given the complexity of material and social interactions on an urban scale, we find that currently there are no concrete definitions upon which these claims can be measured and compared. Therefore, in order to make the ambitious targets of low- and zero-carbon emissions meaningful concepts in the context of urban planning, a carbon accounting framework needs to be rigorously defined and adapted to the urban scale (5259-60).

Determining the extent of the 'urban scale' in terms of emission boundaries is a challenging and often arduous process as boundaries can extend almost indefinitely. Moreover, the intricate and continually changing nature of cities, urban areas, precincts and the systems sustaining them, as well as the activities going on within them, make carbon calculations extremely complex (Brugmann 2010; Kennedy and Sgouridis 2011). Nevertheless, several attempts and proposals have been made to define these boundaries and calculate the carbon footprint of entire cities and urban areas (C40/ICLEI/WRI 2012; ICLEI 2009; Kennedy and Sgouridis 2011; Ramaswami et al. 2008). These will be discussed in Chap. 8. An overview of the key issues associated with carbon footprinting/GHG accounting highlighted in Sect. 7.2 is provided in Appendix A, along with the methodologies currently underpinning the practice.

---

[8] Scope 1, 2 and 3 emissions are defined by the Kyoto Protocol regulations and GHG Protocol as discussed later.

Table 7.2 Key considerations for carbon footprint and definitions

Issue	Description
Gases	Which gases are included? For example, only carbon dioxide (CO_2) or multiple greenhouse gases (CO_2-e)?
Measurement units	What unit is the carbon footprint measured in? Is it in weight (e.g. tonnes) or area (e.g. global hectares)?
Accounting methodology	Is an inventory approach or a life cycle analysis (LCA) used? If LCA is chosen, which methodology, i.e. process analysis (PA), environmental input–output (IEO) analysis, hybrid EIO-LCA?
Boundaries	What are the overall system boundaries? i.e. Does it include only operational emissions or are supply chain emissions, such as the embodied emissions in materials included?
Scopes	Which scopes of emissions are included? Direct and indirect? Scope 1, 2 and 3?
Time frame	If LCA is used, what is the lifetime given to the embodied emissions in buildings and infrastructure?
Emissions reduction	Are there targets set for emissions reduction? What are the allowable emissions reduction options?
Offsets	Are offset included? If so, is there a limit? What offsets are eligible? Are they certified, credible offsets?

Adopted from ASBEC (2011), East (2008), Kennedy and Sgouridis (2011), Wiedmann and Minx (2008), WRI and WBCSD (2004)

7.8 Defining the Carbon Footprint

Calculating the carbon footprint and establishing an emissions baseline is the first step in the process of achieving carbon neutrality and should be the basis for any carbon claim. To begin with, however, it is important to note the broad difference between a carbon footprint and GHG accounting.

A carbon footprint generally utilises a life cycle approach to calculate emissions and is most often applied to products. GHG accounting, on the other hand, focuses more on assessing the operational GHG emissions and is commonly represented in the form of a GHG inventory. The inventory approach is usually applied to organisations, services or individuals (Bolwig and Gibbon 2009) and is predominantly based on the GHG Protocol (WRI and WBCSD 2004). It is this approach that is most often applied to cities and local governments.

When calculating the emissions associated with a precinct-scale development, a carbon footprint approach appears to be a more appropriate choice, as it allows greater attention to be given to the physical aspects of the built environment as opposed to only focusing on the business activities within its boundaries. It can thus provide a far more detailed analysis of the precinct, including the choice of materials and resource management systems, thereby encouraging additional GHG abatement opportunities to be sought and thus stimulated throughout the economy. Buildings and infrastructure should not be seen in isolation; the operational emissions associated with buildings, transport and resource management systems need to also be taken into account. Therefore, a combination of the two approaches is required. Some of the models being proposed that assess carbon within precincts are starting to include both of these emissions approaches (Beattie et al. 2012).

Nevertheless, the term 'carbon footprint', similar to the term 'carbon neutrality', is rarely adequately, rigorously or universally defined (Lean and Kay 2008; Struck 2010; Wiedmann and Minx 2008). Wiedmann and Minx (2008) identify and discuss numerous issues that should be considered and clearly communicated when defining a carbon footprint or a GHG inventory. These issues, along with other key issues identified in the literature (ASBEC 2011; East 2008; UK Department of Communities and Local Government 2008, 2009; Kennedy and Sgouridis 2011; WRI and WBCSD 2004), are provided in Table 7.2 and discussed in more detail, and specifically in the context of the built environment and urban precinct development, in Appendix A.

The degree to which carbon footprints and inventories vary depending on the inputs and decisions made in relation to the issues outlined above can be substantial. These issues need to be sufficiently addressed and clearly articulated within the definition and/or description of the process of how a carbon goal is achieved.

7.9 The Offset Polemic

As discussed at the beginning of this chapter, the concept of carbon offsetting has grown considerably over the last decade, leaving in its wake a thriving voluntary carbon offset market. However, there have been concerns and issues along the way, particularly relating to transparency and credibility resulting from the initial unregulated nature of the voluntary carbon market (Lovell et al. 2009; Moore 2009; Murray and Dey 2008). However, this market has rapidly matured with radically improved processes, structures, standards and regulations now underpinning it (Peters-Stanley et al. 2011). The mandatory offset market, represented by offsets generated under Kyoto Protocol's Clean Development Mechanism and Joint Implementation, has had a significant degree of regulation and stringency since their inception. Nevertheless, there are still many who oppose offsetting, contesting the fundamental principle and concept of it. These issues and others are discussed below.

7.9.1 The Perceived Fundamental Flaw

The rapid development of the voluntary offset market clearly demonstrates the appeal of carbon offsetting by companies and individuals as a way of exhibiting environmental commitment and action. However, not everyone remains convinced of the virtues of it. Smith (2007) relates the concept to the story of the pardoners in the Middle Ages who sold the benefit of their good deeds to sinners who could purge themselves of their indulgences for a fee, allowing sinners to continue transgressing. More recently, a parody of carbon offsetting was made using cheating as the commodity (see cheatneutral.com), which allows one couple to offset their infidelities using another couple's fidelity and faithfulness. Murray and Dey (2008) point out that in both situations, there is no net reduction of the sinning, transgressions or heartache, and there is only a balancing of the commodities.

Friends of the Earth and other NGOs also stand deeply opposed to the concept of offsetting, arguing that it causes more harm than good (see Bullock et al. 2009). They contend that it will not lead to overall emissions reductions, suggesting it merely legitimises the idea that people who can afford to pay others to reduce their emissions can do so and thus continue to pollute as normal. They also suggest that it delays investment in essential infrastructure, particularly in developed countries where investment may have gone ahead regardless, but now waits to determine if a project is eligible to generate offsets. Finally, they argue that carbon offsetting institutionalises the idea that emissions reductions can occur in either developed ***or*** developing countries, ignoring the scientific recommendation that reductions are urgently required in all locations (Bullock et al. 2009).

However, as carbon offsetting is based on the premise that emissions have the same effect on the atmosphere wherever they are produced or abated in the world, it is argued, particularly by economists, that the least cost carbon abatement opportunities should always be pursued first, regardless of their location. Thus, if it is cheaper to reduce emissions offsite through offsets than through onsite emissions reduction measures, the cheaper option should always be the preferred choice. It naturally follows that the price of offsets will increase over time as offset projects become scarcer (e.g. tree planting) or projects become ineligible (e.g. if replacing light bulbs in India becomes the norm and no longer considered 'additional'), making domestic, onsite carbon abatement more financially viable.

Both arguments are valid and will no doubt continue to be contested. Most carbon-neutral schemes, however, do strongly recommend that in-house emissions reduction options be pursued prior to offsetting.

Lovell et al. (2009) also analyse the concept of offsetting in relation to sustainable and ethical consumption. They examine both the notion of nature as a commodity, which looks at the implications of marketing 'the environment', as well as the effect offsets have on driving behaviour and consumer's choice of products. Their analysis proposes that the voluntary offset market is indeed a complex and uncertain one, but that ultimately, it can help to shape behaviours by making people aware of the effect their everyday, often mundane actions, such as driving a car, has on the environment and atmosphere.

Finally, it is important to see that some offsets, such as biomass plantings (e.g. CDM projects in developing countries and other voluntary offsets in developed countries), can have substantial co-benefits such as increasing local biodiversity and reducing land degradation (Karousakis 2009). Many of these important environmental projects would not proceed without the income that carbon credits provide. There are many offset projects that link a city with its own bioregion through tree planting (e.g. the Gondwana Link project in Perth[9]) highlighting the importance of bioregional partnerships with cities (see Newman and Jennings 2008). Achieving carbon neutrality can be an important legacy for a lot more than just climate action.

7.9.2 The Voluntary Carbon Market—Integrity of Offsets

While offsets generated for use in the compliance market have been relatively well regulated from the beginning, Murray and Dey (2008) expressed concern about the rapid expansion of the voluntary carbon market. They questioned the integrity of offsets generated in this market, suggesting there was insufficient infrastructure, standards and frameworks in place to deal with a range of issues such as additionality, eligibility requirements of offsets (e.g. how long should a tree be in the ground), carbon accounting methodologies (i.e. how to calculate the emission abatement potential of specific projects) and how to trade voluntary offsets, with the assurance that offsets are not sold twice. Fortunately, since expressing this original concern in 2006, most of these issues have been addressed with a relatively high degree of rigour (Peters-Stanley et al. 2011). There are increasingly stringent standards now underpinning the voluntary carbon market, and new internationally peer-reviewed methodologies are constantly being developed for a range of carbon offset projects. Furthermore, a growing number of offset providers (over 66 % in 2010) are using reputable carbon registries to retire their transacted carbon credits in order to demonstrate greater transparency and credibility (Peters-Stanley et al. 2011).

Another option for voluntary market participants proposed by Kollmuss and Lazarus (2010) involves the voluntary purchasing of tradable 'allowance units' from compliance markets (i.e. government administered emission allowances that are distributed under a cap and trade scheme), instead of offsets generated in the voluntary market. Allowing voluntary purchases of compliance units avoids issues of double counting and other integrity issues often associated with voluntary offsets (Bullock et al. 2009; Moore 2009), as well as having the added benefit of compelling covered sectors to undertake greater in-house emissions reductions since the

---

[9] See http://www.gondwanalink.org/index.aspx.

overall amount of units available to the compliance market would be reduced (Kollmuss and Lazarus 2010).

Moore (2009) discusses the key issues surrounding voluntary carbon offsets in depth and clearly lays the foundation on why a standard needed to be introduced. In 2010, the Australian Federal Government introduced the National Carbon Offset Standard, which effectively addressed the issues highlighted in her paper. This standard is discussed in Chap. 11. The voluntary offset market can thus be confidently seen as a scientifically credible process. However, the opportunity for urban developments to claim credits for their emissions reductions directly in a mainstream carbon market is not yet there.

7.10 Conclusion

This chapter examined the rise in popularity of the concept of carbon neutrality, which was largely a result of the sudden emergence and growth in the voluntary carbon offset market.

A review of carbon-neutral definitions highlighted a common understanding of the term and broad consensus over the general process (i.e. measure, reduce, offset/neutralise), and few definitions explained in detail how carbon neutrality should be achieved. This was particularly evident in the case of the built environment and precinct development, as most carbon-neutral definitions were tailored more to businesses and individuals.

A variety of factors that need to be considered when undertaking a carbon analysis were identified, which, without a consistent approach in addressing them, could result in very different carbon outcomes for the same project. The main issues requiring consideration specifically in the context of precinct-scale urban development were discussed, along with some current issues associated with the generation of offsets for the urban development sector.

The issues raised in this chapter made it clear that a consistent approach and framework for determining the GHG emissions associated with urban development is required. Importantly, it also demonstrated the need for a specific process that can acknowledge the unique opportunities for innovative onsite carbon reductions that exist at the urban precinct level.

References

Australian Institute of Landscape Architects (AILA). (2010). *Climate adaptation tools for sustainable settlements (CATSS)*. Canberra: Australian Institute of Landscape Architects.

Australian Sustainable Built Environment Council (ASBEC). (2011). *Defining zero emission buildings. Review and recommendations: Final Report*. Retrieved from http://www.asbec.asn.au/files/ASBEC_Zero_Carbon_Definitions_Final_Report_Release_Version_15112011_0.pdf.

BBC News. (2006, December 13). *Zero Carbon' homes plan unveiled.* Retrieved from http://news.bbc.co.uk/2/hi/science/nature/6176229.stm.

Beattie, C., Bunning, J., Stewart, J., Newman, P., & Anda, M. (2012). Measuring carbon for urban development planning. *The International Journal of Climate Change: Impacts and Responses, 3*(4), 35–52.

Boake, T. M. (2008). The leap to zero carbon and zero emissions: Understanding how to go beyond existing sustainable design protocols. *Journal of Green Building, 3*(4), 64–77.

Bolwig, S., & Gibbon, P. (2009). *Counting carbon in the marketplace: Part 1-overview paper: Report for the OECD.* Trade and Agriculture Directorate; Joint Working Party on Trade and Environment.

Brugmann, J. (2010). *Welcome to the urban revolution: How cities are changing the world.* New York: Bloomsbury Press.

Bullock, S., Childs, M., & Picken, T. (2009). *A dangerous distraction. Why offsetting is failing the climate and people: The evidence.* London: Friends of the Earth.

C40/ICLEI/WRI. (2012). *Global protocol for community-scale GHG emissions. Version 0.9.* Retrieved from http://siteresources.worldbank.org/INTUWM/Resources/340232-1332259778466/GPC_v9_20120320.pdf.

Carbon Neutral (UK private Company). (n.d.). *Carbon Glossary.* Retrieved from http://www.carbonneutral.com/resource-hub/carbon-glossary.

Department of Climate Change and Energy Efficiency (DCCEE). (2011a). *Greenhouse friendly.* Retrieved from http://www.climatechange.gov.au/greenhousefriendly/.

Department of Climate Change and Energy Efficiency (DCCEE). (2011b). *Carbon neutral program guidelines: National carbon offset standard. Version 2.0.* Retrieved from http://www.climatechange.gov.au/sites/climatechange/files/documents/03_2013/cnp-guidelines.pdf.

East, A. J. (2008). What is a carbon Footprint? An overview of definitions and methodologies. Vegetable industry carbon footprint scoping study. Sydney.

Eichholtz, P., Kok, N., & Quigley, J. (2008). Doing well by doing good? Green Office Buildings. Program on Housing and Urban Policy, Working Paper No. W08-001. Institute of Business and Economic Research, Fisher Center for Real Estate and Urban Economics.

Ewing, R., Bartholomew, K., Winkelman, S., Walters, J., & Chen, D. (2008). *Growing cooler: The evidence on urban development and climate change.* Washington, DC: Urban Land Institute.

Fisk, W., & Seppanen, O. (2007). *Providing better indoor environmental quality brings economic benefits.* Paper presented at the REHVA World Congress Clima 2007 Wellbeing Indoors, Helsinki, Finland.

Garnaut, R. (2008). *The Garnaut climate change review-final report.* Melbourne: Cambridge University Press.

Gore, A. (2006). *An inconvenient truth: The planetary emergency of global warming and what we can do about it.* New York: Melcher Media.

Hall, S. (2014). Green building performance evaluation for existing commercial buildings (Doctoral Dissertation, Curtin University, Curtin University Sustainability Policy Institute, 2014).

ICLEI. (2009). International local government GHG emissions analysis protocol (IEAP). Available online http://carbonn.org/fileadmin/user_upload/carbonn/Standards/IEAP_October2010_color.pdf: ICLEI—Local Governments for Sustainability.

Intergovernmental Panel on Climate Change (IPCC). (2007). Climate change 2007: Mitigation of climate change. In B. Metz, O. Davidson, P. Bosch, R. Dave, & L. Meyer (Eds.), *Contribution of working group III to the fourth assessment report of the intergovernmental panel on climate change.* Cambridge: Cambridge University Press.

Karousakis, K. (2009). Promoting biodiversity co-benefits in REDD. OECD Environment Working Papers, No. 11. Retrieved from http://www.cbd.int/climate/doc/oecd-promoting-biodiversity-redd-en.pdf.

Kennedy, S., & Sgouridis, S. (2011). Rigorous classification and carbon accounting principles for low and Zero Carbon Cities. *Energy Policy* 5259–5268.

Kollmuss, A., & Lazarus, M. (2010). Buying and cancelling allowances as an alternative for the voluntary offset market: A preliminary review of issues and options. Working Paper, Stockholm Environment Institute.

Kubba, S. (2012). *Handbook of green building design and construction*. Oxford: Elsevier Science & Technology.

Lean, G., & Kay, B. (2008, March 30). Four nations in race to be first to go carbon neutral. The Independent. Retrieved from http://www.independent.co.uk/environment/climate-change/four-nations-in-race-to-be-first-to-go-carbon-neutral-802627.html.

Lehmann, S. (2015). *Low carbon cities: Transforming urban systems*. New York: Routledge.

Lovell, H., Bulkeley, H., & Liverman, D. (2009). Carbon offsetting: Sustaining consumption? *Environment and Planning, 41*, 2357–2379.

Low Carbon Australia. (2011). Carbon Neutral Program Guidelines. Retrieved from http://www.climatechange.gov.au/government/initiatives/low-carbon-australia/ncos-carbon-neutral-program/carbon-neutral-program-guidelines.aspx.

Moore, L. (2009). Voluntary carbon offsets: A legal perspective. In W. Gumley & T. Daya-Winderbottom (Eds.), *Climate change law: Comparative, contractual and regulatory considerations*. Germany: Lawbook Company.

Murray, J., & Dey, C. (2008). The carbon neutral free for all. *International Journal of Greenhouse Gas Control, 3*, 237–248.

Newman, P., & Jennings, I. (2008). *Cities as sustainable ecosystems: Principles and practices*. Washington, DC: Island Press.

Newman, P., Beatley, T., & Boyer, H. (2009). *Resilient cities: Responding to peak oil and climate change*. Washington, DC: Island Press.

Oxford Dictionary. (n.d.). Definition of carbon neutral. Retrieved from http://oxforddictionaries.com/definition/english/carbon-neutral?q=carbon+neutral.

Peters-Stanley, M., Hamilton, K. Marcello, T., & Sjardin, M. (2011). Back to the future: State of the voluntary carbon markets 2011. Bloomberg New Energy Finance, Ecosystem Marketplace.

Ramaswami, A., Hillman, T., Janson, B., Reiner, M., & Thomas, G. (2008). A demand-centered, hybrid life-cycle methodology for city-scale greenhouse gas inventories. *Environmental Science and Technology, 42*(17), 6455–6461.

Rauch, J. N., & Newman, J. (2007). Defining sustainability metric targets in an institutional setting. Retrieved from http://www.emeraldinsight.com/doi/abs/10.1108/14676370910945927.

Roseland, M. (2012). *Toward sustainable communities: Solutions for citizens and their governments* (4th ed.). Gabriola Island, BC: New Society Publishers.

Rosenthal, E. (2007). Vatican penance: Forgive us our carbon output. *The New York Times*. Retrieved from http://www.nytimes.com/2007/09/17/world/europe/17carbon.html?_r=2.

Roussac, C., & Kuiper, G. (2009). How low can we go? What is the scope for reducing emissions from commercial real estate? Working Paper No. 1, Investa Sustainability Institute.

Smith, K. (2007). *The carbon neutral myth: Offset indulgencies for your climate sins*. Carbon Trade Watch, Amsterdam: Transnational Institute.

Stern, N. (2006). *The economics of climate change: The Stern review*. Cambridge: Cambridge University Press.

Struck, D. (2010, April 10). Carbon offsets: How a Vatican forest failed to reduce global warming. *The Christian Science Monitor*. Retrieved from http://www.csmonitor.com/Environment/2010/0420/Carbon-offsets-How-a-Vatican-forest-failed-to-reduce-global-warming.

Sturgis, S., & Roberts, G. (2010). *Redefining Zero: Carbon Profiling as a solution to whole life carbon emissions measurement in buildings*. London: RICS Research.

The Climate Group. (2010). Climate smart precincts: Opportunities report to the first ministers the climate group.

UK Department for Communities and Local Government. (2007). *Building a greener future: Policy statement*. HM Government. Retrieved from http://www.rbkc.gov.uk/PDF/80%20Building%20a%20Greener%20Future%20Policy%20Statement%20July%202007.pdf.

UK Department of Communities and Local Government. (2009). *Guidance on carbon neutrality*. London: CLG. Retrieved from http://webarchive.nationalarchives.gov.uk/20110508074721/http://decc.gov.uk/assets/decc/What%20we%20do/A%20low%20carbon%20UK/carbonneutrality/1_20090930090921_e_@@_carbonneutralityguidance.pdf.
Wiedmann, T., & Minx, J. (2008). A definition of "Carbon Footprint". *Ecological Economic Research Trends*. NY, USA: Nova Science Publishers.
Wikipedia. (n.d.). Carbon neutral. Retrieved from http://en.wikipedia.org/wiki/Carbon_neutral.
World Resources Institute and World Business Council for Sustainable Development (WRI and WBCSD). (2004). The greenhouse gas protocol: A corporate accounting and reporting standard, revised edition. Retrieved from http://www.ghgprotocol.org/files/ghgp/public/ghg-protocol-revised.pdf.
Zuo, J., Read, B., Pullen, S., & Shi, Q. (2012). Achieving carbon neutrality in commercial building developments—Perceptions of the construction industry. *Habitat International, 36*, 278–286.

Chapter 8
Counting Carbon in Cities

8.1 Introduction

Determining the carbon footprint or greenhouse gas emissions associated with a product or organisation can be a complex undertaking, as demonstrated by the variety of issues discussed in Chap. 7. However, addressing these issues at a precinct, community or a citywide scale is considerably more difficult task as the magnitude of each issue increases. Nevertheless, despite this challenge, numerous attempts have been made by academia, industry, cities and local communities to classify these emissions in order to better understand the GHG contribution of urban development at the various levels.

The following is a discussion of key literature pertaining to GHG accounting at the city and precinct scales. Several prominent existing and proposed city-scale GHG methodologies, tools and inventories are highlighted, and a brief description of the most recognised and utilised of these frameworks and initiatives is provided. This is followed by a summary of existing tools that target GHG emissions at the precinct scale. Finally, a discussion of the suitability of the tools, for the purpose outlined in this book, is provided.

8.2 Determining the GHG Contribution of Cities

As previously noted, cities and urban areas are increasingly being identified as major producers of global GHG emissions (Dhakal 2010; Dodman 2009; Hoornweg et al. 2011; Satterthwaite 2008; World Bank 2010). However, the perceived extent of their contribution varies depending on which emissions are included in the GHG analysis of any given city. For example, Satterthwaite (2008) contends that a significant proportion of the emissions that are often attributed to cities actually occur outside a city's official legislative boundaries (e.g. from fossil fuel power stations, waste

V. Rauland and P. Newman, *Decarbonising Cities*, Green Energy and Technology, DOI 10.1007/978-3-319-15506-7_8

management or agricultural practices), but occur as a result of a city's demand for resources or their urban activities. A question that therefore arises is: should emissions be assigned to the location of their production or at the place of their consumption?

The issue of boundaries thus forms a major component of the discussion around city GHG inventories. Assigning emissions to consumption acknowledges the fact that emissions generated in one city or country (particularly the case for developing nations) are often produced primarily to satisfy demand for commodities in other cities or countries (particularly industrialised nations) (Dhakal 2010; Hoornweg et al. 2011; Satterthwaite 2008). Nevertheless, accounting for consumption is particularly tricky and thus most GHG frameworks have tended to focus predominantly on the production side.

The concept of urban metabolism, first introduced by Wolman (1965), clearly explains how resources flow in and out of cities, and the subsequent environmental impacts associated with them. Newman and Kenworthy (1999) applied this approach to an Extended Metabolism Model where the reduction in resource inputs and waste outputs as well as improved liveability in cities was used to explain what sustainability meant to cities. Later research led to the development of a useful tool based around the flow of materials (Materials Flow Analysis, or MFA). This could potentially provide an appropriate methodology for determining emissions associated with cities. However, as the analysis often uses a region's physical boundary, Kennedy and Sgouridis (2011) note that it is not suitable for cities and urban precincts, which are usually dependent on several emission sources outside their boundary (i.e. electricity production), but that which they are directly responsible for. Ecological footprinting, which was developed by Rees and Wackernagle (1996), included resource inputs and waste outputs on the basis of a single comparative unit of land to dramatise the impact of cities on their bioregions and the planet. This can motivate action to reduce carbon but cannot help in the detailed process of recognising how to decarbonise urban development in a scientific way.

Numerous GHG accounting frameworks, methodologies and inventories have been developed in recent years, and countless more have been proposed. These have attempted to address a variety of urban scales, including communities, local governments and cities, as each level acknowledges and accepts the importance of their role in addressing climate change (Carbon Disclosure Project 2012; Hoornweg et al. 2011).

The GHG frameworks and inventories are being developed and/or proposed by a range of entities including cities and individual communities,[1] NGOs, municipal associations and organisations (e.g. ICLEI Local Governments for Sustainability, World Bank and C40 Cities), as well as the private sector—often in the form of industry partnerships (e.g. ARUP and C40 cities, and CDP Cities and AECOM) and

[1] See the report 'Measurement for Management: CDP Cities 2012 Global Report' (Carbon Disclosure Project 2012) for an analysis of the various inventories that participating cities use to disclose emissions.

academia (Dhakal 2010; Dixit et al. 2012; Dodman 2009; Hoornweg et al. 2011; Satterthwaite 2008; Sovacool and Brown 2009; World Bank 2010; Xi et al. 2011).

Most frameworks and inventories have drawn from previous experience and earlier institutional standards and protocols. Some key existing documents include the International Local Government GHG Emissions Analysis Protocol (IEAP) developed by ICLEI, the Draft International Standard for Determining Greenhouse Gas Emissions for Cities, developed in partnership between UNEP/UN and HABITAT/WB, the GHG Protocol Standards (WRI/WBCSD), the Baseline Emissions Inventory/Monitoring Emissions Inventory methodology (EC-CoM JRC) and the Local Government Operations Protocol produced by ICLEI-USA (C40/ICLEI/WRI 2012).

Twelve of the most prominent existing and proposed GHG methodologies, tools and inventories for urban areas and cities are listed in date order in Table 8.1. Additional tools and inventory approaches have been analysed in other studies (Bader and Bleischwitz 2009; ICLEI 2010; Sugar et al. n.d.), but these are not relevant to this book. A brief description of the five most recognised and utilised of these 12 international and Australian frameworks, platforms and initiatives relevant to the city scale is provided below Table 8.1.

8.3 Frameworks and Initiatives for Calculating City-based Emissions

8.3.1 ICLEI—Local Governments for Sustainability

ICLEI—Local Governments for Sustainability was one of the first organisations to draw attention to the GHG abatement opportunities available at the local level (Sugar et al. n.d.). ICLEI was formed in 1990 as the International Council for Local Environmental Initiatives and was the outcome of a summit hosted at the United Nations in New York by the United Nations Environment Programme (UNEP) that brought together more than 200 local governments to discuss how they could best promote and implement environmental action at the local level. On this occasion, the newly established organisation also launched the Local Agenda 21 programme initiative, which was subsequently endorsed by the UN Conference on Environment and Development in 1992. This local government association rapidly grew as governments and councils around the world realised the potential for knowledge sharing and capacity building around local environmental programmes and initiatives. By 1998, local governments from more than 240 cities around the world had committed to reporting their greenhouse gas emissions and putting in place management plans to reduce emissions as part of a global campaign called ‘Cities for Climate Protection’ (CCP) (Sugar et al. n.d).

In 2003, ICLEI members elected to change the name to better reflect the association’s broader focus on environmental and social issues and thus ICLEI became

Table 8.1 Primary tools and frameworks for GHG measurement relevant to cities. Adapted from ICLEI (2010)

Name of tool/protocol/framework	Country focus	Organisation responsible	Framework/tool focus	Released
GHG protocol standards	International	WRI/WBCSD	Corporate and project level	Original: 2001 updated: 2004/5
C40 cities climate leadership group	International	C40 cities and Clinton climate initiative	Mega-cities	2006
Resources and energy analysis programme (REAP)	UK	Stockholm environment institute	Consumption-based emissions at the local, region and national levels	2006
Local government operations protocol	USA	ICLEI-USA	Local government	Original: 2008 updated: 2010
Low-carbon cities programme (LCCP)	UK	UK Carbon Trust	Citywide (public and private sector and community)	2008 (project only ran for 6 months)
International local government GHG emissions analysis protocol (IEAP)	International	ICLEI—Local governments for sustainability	Local government	2009
Carbon neutrality framework for local government—Australian version	Australia	ICLEI oceania	Local government and/or community	2009
WALGA GHG reporting platform	Australia	Western Australia local government association and Greensense	Local government	2009
Draft international standard for determining greenhouse gas emissions for cities	International	UNEP/UN-HABITAT/World Bank	Cities	2010
carbon*n* Cities Climate Registry (cCCR)	International	The Bonn Center for Local Climate action and reporting	Local governments	2010
CDP Cities GHG reporting platform	International	CDP cities, C40 cities, Clinton climate initiative, Aecom	Cities	2010
Global protocol for community-scale GHG emission (GPC)	International	C40 cities, ICLEI, WRI, World Bank, UNEP, UN-HABITAT	Local government and community	2014

'ICLEI—Local Governments for Sustainability'. There are now over 1220 members of ICLEI representing local governments from 70 countries worldwide that are committed to achieving sustainable development (ICLEI—Local Governments for Sustainability 2012a).

8.3.1.1 History

In 1993, in partnership with UNEP, ICLEI convened the first Local Government Leaders' Summit on climate change at the United Nations in New York. As a follow-up to this summit, ICLEI convened a second summit at the first meeting of the Conference of the Parties (COP) to the United Nations Framework Convention on Climate Change (UNFCCC) in Berlin in 1995. At this meeting, ICLEI was invited to send a small delegation of mayors to speak to the government negotiators regarding local actions to reduce GHGs and their economic benefits and impacts. Thereafter, ICLEI began convening summits on the occasion of the COP meetings and ICLEI mayor's delegation was invited to speak to attending ministers and delegates. This arrangement continues to this day.

Over the years, ICLEI has developed several guidelines to assist local governments in preparing GHG reduction plans as well as protocols for GHG baseline studies and for reporting to enable inter-city comparisons and deeper understanding of how these emissions contribute to emissions at national and global levels. To aid the centralised reporting and analysis of local GHG reductions internationally, in 2010 ICLEI establish the Carbon Climate Registry in partnership with the cities of Bonn and Mexico City. Nearly 500 local governments currently report their GHG emissions to the registry using a standardised protocol (The Bonn Center for Local Climate Action and Reporting 2012).

In 2005, ICLEI established the World Mayors Council on Climate Change (WMCCC) with Mayor Yorikane Masumoto, Mayor of Kyoto, on the occasion of the establishment of the Kyoto Protocol. The WMCCC currently has 80 mayor council members and also meets on the occasion of the UN COP meetings.

ICLEI has also collaborated with, and has been a key player in, other international campaigns and initiatives such as the European Campaign for Sustainable Cities and Towns, CDP Cities and Clinton Climate Initiative (CCI)/C40 Cities (discussed below). In 2014, the C40 joined in partnership with ICLEI to further develop and promote the Carbonn Climate Registry.

8.3.1.2 ICLEI—Cities for Climate Protection

CCP was a global campaign established by ICLEI in 1993 (ICLEI 2012b). The CCP was the programmatic offshoot of the first international research project on urban GHG measurement and management, called the Urban CO_2 Reduction Project. The Urban CO_2 Reduction Project was established by ICLEI in 1991 with participating teams from a range of cities around the world. The project established the first of a series of ICLEI software programmes for local level GHG inventories. It also established the '5 Milestone Process' for urban GHG reduction planning that became the cornerstone of the subsequent CCP.

Between 1993 and 2000, the CCP expanded into a number of regional and national scale programmes funded by national governments and international development agencies. CCP programmes have been operated in Australia, NZ, Brazil, Canada, Europe, India, Indonesia, Mexico, the Philippines, South Africa and the USA. By 2000, more than 500 local governments in more than 30 countries were engaged in the CCP programme, applying the 5 Milestone Frameworks.

The establishment of the C40 cities in 2005 brought additional organisational resources to bear in supporting many of the world's larger cities. In Australia, the CCP campaign was managed through a partnership between ICLEI Oceania, the Australian Federal Government's Department of the Environment, Water, Heritage and the Arts, and local councils. By 2008, over 184 Australian councils were participating in the programme and had abated over 18 million tonnes CO_2-e and thus saved themselves and their communities over $95 million in energy costs (ICLEI 2008a). In 2009, however, the Australian Federal Government withdrew its funding support for the programme resulting in its abrupt cessation. While ICLEI still has a presence in Australia and continues to produce documents to assist councils, such as the 'Carbon Neutrality Framework for Local Government' (ICLEI 2008b), ongoing hands-on support for councils is no longer available, nor is the Australian CCP programme data collected or disseminated.

Nevertheless, to ensure that councils continued their progress, new innovative local government GHG reporting platforms and support have emerged to replace ICLEI's CCP programme. One such platform is the Western Australian Local Government Association (WALGA) 'GHG Reporting Framework', which is supported by local climate change consultancy Greensense™. This online platform and programme allows local councils who are members of WALGA to upload their emissions data, share information with other councils and have access to, and be supported by, a dedicated person at WALGA and Greensense™.

8.3.1.3 ICLEI—International Local Government GHG Emissions Analysis Protocol

In 2009, ICLEI released the International Local Government GHG Emissions Analysis Protocol (IEAP) to bring standardisation to GHG inventories at the local government level. The Protocol was designed to assist not only with local government internal (that is, operational) emissions, but also to provided additional guidance on calculating the emissions from the wider community's activities (ICLEI 2009). Sectors to be included in the inventory comprise energy (stationary, transport and fugitive emissions), industrial processes, agriculture, land-use, land-use change and forestry, and waste both from local government- and community-owned operations such as landfill and waste incineration (ICLEI 2009).

8.3.2 Global Protocol for Community-scale GHG Emissions

After recognising the need for and seeing the advantage in having a more harmonised approach to understanding cities emissions, several key organisations (C40 Cities Climate Leadership Group, ICLEI Local Governments for Sustainability, World Resources Institute, World Bank, UNEP, UN-Habitat, GHG Protocol and industry partners such as ARUP) joined together to develop a new standard for calculating emissions at the local level. Named the 'Global Protocol for Community-Scale GHG Emissions' (GPC), it helps to 'improve policy coordination and dissemination of best practices between local governments facing comparable climate-related challenges, and facilitating climate-linked finance through a verifiable, common standard' (C40/ICLEI/WRI 2012: 1). The standard will provide far greater comparability of emissions between cities and increase the integrity of local government claims. A pilot version was released in 2012 for cities and communities to trial, and in December 2014 at the UN Climate Change talks in Lima, Peru, the final version was formally launched.[2] This tool is now the new international standard under the GHG Protocol for city-based GHG accounting.

8.4 Programmes Facilitating Emission Analysis and Reporting

8.4.1 C40 Cities and the Clinton Climate Initiative

C40 Cities was set up in 2005 by then Mayor of London, Ken Livingstone, to encourage the world's largest cities to begin to reduce their emissions and take action on climate change. The first meeting brought together key figures from 18 global megacities and concluded with an agreement by each city to actively work towards reducing their city's emissions and to create an environment that encourages participants to swap ideas and share experiences. A key focus is on accelerating the adoption of new low-carbon technologies by encouraging partnerships between cities and industries and to create joint procurement policies and alliance-structures for purchasing such technologies to increase general uptake in the marketplace (C40 Cities 2013).

In 2006, The Clinton Foundation launched another international city-focused initiative called the CCI, which took a business-centred approach to engaging cities in the fight against climate change. CCI focused on employing practical, quantifiable and industry-focussed measures and actions at the city level. Seeing the clear link between the two initiatives, CCI soon joined forces with C40 Cities, with CCI acting as the delivery vehicle for C40 Cities. The actions, initiatives and measures

[2] See http://www.ghgprotocol.org/city-accounting.

being undertaken by participating cities[3] to address emissions under the joint initiative are documented in various reports produced by their industry partner Arup (ARUP and C40 Cities Climate Leadership Group 2011). The representative body that was created as a result of this initiative is called the C40 Cities Climate Leadership Group.

Also emerging out of this initiative was a new programme called Carbon Positive Development Programme (CPDP), which has identified 18 urban development projects from participating C40 cities that aimed to go beyond net-zero to create carbon positive developments. The programme provides a platform for the various representatives from the projects to share experiences and ideas, collaborate and work towards achieving this goal (C40 Cities Climate Leadership Group and CCI 2012).

8.4.2 CDP Cities

In 2010, the Carbon Disclosure Project, which had previously focused on getting the 500 largest global companies to measure and report their carbon emissions, took on a new challenge and set about creating a similar platform for cities. The CDP Cities Project partnered with existing initiatives, including the C40 Cities Climate Leadership Group, to encourage currently participating cities to report using this platform, as well as promoting the concept to new cities. As of 2012, 73 cities are reporting their emissions, and 45 of these are participating under the C40's initiative (Carbon Disclosure Project 2012).

Cities that join the project do so voluntarily and can choose what information they disclose, e.g. their current city GHG inventories, GHG reduction targets and reduction actions.[4] The project does not stipulate what cities should include in their inventories or which methodologies to use but simply provides a platform for cities to report and communicate on. The project, such as C40's Cities Climate Leadership Group, allows cities to discuss issues, share learnings and be part of a collaborative, proactive demonstration group that is showing the rest of the world how cities can take the lead in acting on climate change.

The first CDP report, released in 2011, revealed a wide variety of emission reporting styles, demonstrating that there is currently no single, or preferred, GHG accounting methodology among cities. It states that 'the data suggests that many cities recognise the value in international protocols but still find them lacking' (KPMG et al. 2011: 14). Several cities are using existing methodologies, such as ICLEI's local governments GHG reporting framework, or they have combined

[3] In 2012, 59 cities are members of C40 Cities/CCI Initiative and have committed to acting on climate change.

[4] Only 70 % of the cities reported their emission inventories through the CDP Cities Project in 2012 (Carbon Disclosure Project 2012).

existing standards and/or frameworks (e.g. GHG protocol and IPCC) with proprietary tools to create new methodologies that use locally relevant information (Carbon Disclosure Project 2012; KPMG et al. 2011).

The current variability in GHG methodologies utilised by participating cities means the emissions sources included in each city's inventory have the potential to vary significantly. Such inconsistency in reporting methods makes it difficult to compare cities' emission profiles. Nevertheless, it is the intent of the project, and believed by the organisers, that greater consistency in accounting methodologies and reporting methods will occur naturally over time.

The style of reporting also differs between cities. Excel spreadsheets currently appear to dominate as the basis for data collection, although there appears to be a shift towards more sophisticated, analytical software tools and systems (Carbon Disclosure Project [CDP] 2012). An example of this is the aforementioned Western Australian 'Local Government Emissions Reporting Platform' developed by local private industry, Greensense™, in partnership with the WALGA.[5] This GHG accounting system is compliant with Australia's national reporting standards (National Greenhouse and Energy Reporting Act and Guidelines) and the National Carbon Offset Standard—the standard against which carbon neutrality is certified in Australia—to ensure that the tool is as practical and useful as possible for the local context.

Verification is another interesting component of the reporting of city emissions, and it is starting to play a larger role in the CDP Cities Project. Around 37 % of participating CDP cities currently have their GHG inventories verified by a third-party organisation, stating that it 'is becoming fundamental for establishing credibility of data with stakeholders' (KPMG et al. 2011: 35).

8.5 Discourse on City GHG Methodologies and Attribution

While cities are identified as being major contributors to climate change, they are also recognised as being a powerful force in global GHG mitigation (Bartholomew and Ewing 2009; Dhakal 2010; Dhakal and Strestha 2010; Dodman 2009; Hoornweg et al. 2011; Newman et al. 2009; Sattherwaite 2008; Roseland 2012). Much academic literature has been concerned with identifying the activities and resulting emissions attributable to cities, in order to better manage and reduce city-based emissions. While the discussion continues around the specific emission sources and activities (Dhakal 2010; Dodman 2009; Satterthwaite 2008), this literature has helped to inform public debate and contributed significantly to the development of the various city GHG accounting methodologies and inventories currently being used by several of the initiatives outlined above.

---

[5] See http://www.walgaclimatechange.com.au/walga-reporting-platform.htm.

Some of the key challenges to calculating and managing carbon emissions in cities, as noted by Dhakal and Shrestha (2010), include:

> …data and information gaps, developing long-term scenarios, establishing a consistent urban carbon accounting framework, understanding of the urban system dynamics, and interaction of urban activities related to carbon emissions across the multiple system boundaries, formulating appropriate policies, and operationalizing the policy instruments. (4753)

Fortunately, many of the issues outlined above are being addressed by new initiatives including the collaborative development of the GPC mentioned above. As commonly noted in the literature, adopting a standardised GHG accounting approach, such as the GPC, will help to increase the reliability of placed-based comparisons, particularly between international cities, as well as generating a more accurate understanding of the true GHG attribution for cities globally (Dhakal 2010).

One of the most common topics emerging out of the literature on GHG accounting for cities has been around GHG attribution and system boundaries (Dhakal 2010; Dhakal and Shrestha 2010; Dodman 2009; Hoornweg et al. 2011; Matthews et al. 2008; Satterthwaite 2008). Most research identifies the important effect consumption has on a city's emissions profile. This is largely because traditional city GHG accounting has taken a production-based approach that considers the emissions produced physically onsite or within a city's geographic/legislative boundary (Dodman 2009). When adopting this production-based approach, larger cities often come out having smaller per capita footprints. This is demonstrated in an analysis by Dodman (2009), who compared several UK cities and showed that inhabitants of larger cities indeed often had lower per capita emissions than average UK citizens and significantly lower footprints than citizens of smaller rural cities that are likely to have more industry located in their boundaries but fewer people.

The production-based approach fails to acknowledge the fact that a significant proportion of the emissions generated from the production of commodities in one geographical area (often smaller industrial towns) is generally created primarily to satisfy the demand for those commodities in other areas—typically larger cities. Ramaswami et al. (2008) highlight this fact, that 'producer cities' (i.e. those that produce materials such as cement, steel, food and other key urban materials) often get penalised in the production-based approach, while other 'consumer' cities get rewarded for their recycling.

Nevertheless, significant urban research suggests numerous other reasons exist for the lower GHG emissions associated with larger, denser cities. Factors such as dwelling size, transport mode, infrastructure options and consumption habits can contribute significantly to larger cities having lower per capita carbon footprints compared to their suburban and rural counterparts (Beattie and Newman 2011; Dhakal 2010; Dodman 2009; Ewing et al. 2008; Glaeser 2011; Glaeser and Kahn 2008; Newman et al. 2009; Newman and Kenworthy 1999, 2015; Newton 2008; Rauland and Newman 2011; Roseland 2012). Density also provides economies of scale to make various resource management technologies such as recycling and wastewater treatment more viable, as well as public transport infrastructure (Newman 2014).

Although the consumption-based approach for GHG accounting provides a more holistic and arguably more accurate representation of a city's emissions profile, research into this area is currently limited and there remains no accepted framework for apportioning these consumption-based emissions at a citywide scale (Dhakal and Shrestha 2010).

In their article 'Rigorous classification and carbon accounting principles for low and Zero Carbon Cities', Kennedy and Sgouridis (2011) attempt to include the emissions from both production and consumption within cities and thus provide a useful and comprehensive analysis of citywide emissions. A summary of their proposed scopes and boundaries for the urban scale is provided below in Fig. 8.1.

As can be seen from their illustrations, the urban-scale climate change and sustainability assessment tool developed by Kennedy and Sgouridis (2011) is a relatively complex analysis of urban emissions that take into account all three scopes of emissions. They argue that the broad and relatively simplistic carbon footprint definitions and principles that are often applied to products and organisations are not necessarily appropriate when applied to cities and precinct developments, as these larger areas need to take into account and include multiple stakeholders, a variety of urban scales and numerous emission sources (Kennedy and Sgouridis 2011). However, they also point out that using a definition that is too narrow, such as the one proposed by Wiedmann and Minx (2008), can end up 'underestimating the value of urban efforts for system-wide mitigation' (Kennedy and Sgouridis 2011: 5262) such as those discussed in Chap. 4.

The GHG accounting proposed by Kennedy and Sgouridis (2011) helps to bring a more thorough understanding of the total carbon associated with cities and urban areas and demonstrates how interrelated and complex cities are, and hence how complicated such GHG accounting can be. However, when applied to the precinct

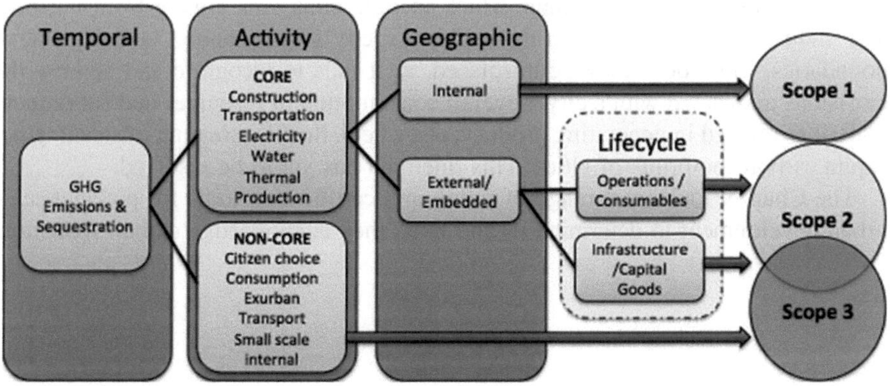

Fig. 8.1 Urban GHG emissions: scoping and boundaries. *Source* Adapted from Kennedy and Sgouridis (2011)

scale, this analysis arguably goes beyond what a developer could legitimately be held responsible for, even in conjunction with other key stakeholders such as utilities. It also runs the risk of being too complicated and time-intensive to get developers willing to participate in calculating and managing such a broad range of emission sources.[6]

Nevertheless, having a comprehensive GHG analysis can be very helpful in developing appropriate carbon policies and mitigation strategies for both developers and local councils, such as sourcing commodities with lower carbon footprints and requiring lower carbon transport infrastructure (Dhakal and Shrestha 2010). It could also help to provide a basis and framework for knowledge sharing, because, as Sovacool and Brown (2009) note, 'the lack of comparative analysis between metropolitan areas makes it difficult to confirm or refute best practices and policies' (4857).

8.6 Conclusion

There is clearly significant research being undertaken and considerable effort being put into developing better and more consistent GHG accounting practices and rating tools for cities, precincts and urban developments. Numerous climate change and GHG measurement tools, initiatives and programmes have been developed and offered to cities and communities over the years by various key international bodies such as ICLEI, UNEP, UN-Habitat, World Bank, WRI and WBCSD, as well as other joint initiatives such as The CCI and C40 cities. The abundance of initiatives and the significant overlap between them led to the amalgamation of several parties and programmes, as well as agreement to collaborate on developing an overarching GHG framework and GHG accounting standard for communities and cities, called the Global Protocol for Community-Scale GHG Emissions (GPC).

However, currently, most city GHG accounting frameworks focus on production-based, operational emissions, which include emissions arising from business and industry activities occurring within a city's geographic and legislative boundaries. This focus may be misplaced, as it fails to recognise and address the emissions associated with a city's overall consumption of resources and subsequent emissions created in generating products elsewhere, thereby creating misleading per capita carbon footprints of cities. This dilemma has yet to be resolved.

The Chap. 9 looks at a range of rating and certification tools for precinct-scale urban development to determine if, and how, they acknowledge carbon reduction.

[6] It is important to note that calculations of different sources of emissions are necessary for different purposes, e.g. for developers, for behaviour change programs for consumers, for town planners or transport planners.

References

ARUP & C40 Cities Climate Leadership Group. (2011). *Climate action in megacities: C40 Cities baseline and opportunities*. Retrieved from http://www.arup.com/Publications/Climate_Action_in_Megacities.aspx.

Bader, N., & Bleischwitz, R. (2009). Study report: Comparative analysis of local GHG inventory tools. In I. V. Environment (Ed.). Retrieved from http://www.institut.veolia.org/fileadmin/medias/documents/491_Final-report-Comparative-Analysis-of.pdf.

Bartholomew, K., & Ewing, R. (2009). Land use-transportation scenario planning: A meta-analysis. *Journal of the American Planning Association, 75*(1), 1–15.

Beattie, C. & Newman, P. (2011). The density trade-off: Does high rise living contribute more than living in detached dwellings to greenhouse gas emissions? In C. Whitzman, & R. Fincher, (Ed.), *State of Australian cities national conference, November 29–December 2 2011*. Melbourne, Vic: Australian Sustainable Cities and Regions Network.

C40 Cities Climate Leadership Group & Clinton Climate Initiative. (2012). *Climate positive development learning program second phase commences in Melbourne*. (Web log post). Retrieved from http://live.c40cities.org/blog/2012/3/26/climate-positive-development-learning-program-second-phase-c.html.

C40 Cities. (2013). History of the C40. *C40 Cities: Climate leadership group website*. Retrieved from http://www.c40cities.org/history.

C40/ICLEI/WRI. (2012). *Global protocol for community-scale GHG emissions. Version 0.9*. Retrieved from http://siteresources.worldbank.org/INTUWM/Resources/340232-1332259778466/GPC_v9_20120320.pdf.

Carbon Disclosure Project [CDP]. (2012). *Measurement for management: CDP Cities 2012 global report*. Retrieved from http://www.cdproject.net/CDPResults/CDP-Cities-2012-Global-Report.pdf.

Dhakal, S. (2010). GHG emissions from urbanisation and opportunities for urban carbon mitigation. *Current Opinion in Environmental Sustainability, 2*, 277–283.

Dhakal, S., & Shrestha, R. M. (2010). Editorial: Bridging the research gaps for carbon emissions and their management in cities. *Energy Policy, 38*, 4753–4755.

Dixit, M. K., Fernández-Solís, J. L., Lavy, S., & Culp, C. H. (2012). Need for an embodied energy measurement protocol for buildings: A review paper. *Renewable and Sustainable Energy Reviews, 16*, 3730–3743.

Dodman, D. (2009). Blaming cities for climate change? An analysis of urban greenhouse gas emissions inventories. *Environment and Urbanization, 21*, 185–201.

Ewing, R., Bartholomew, K., Winkelman, S., Walters, J., & Chen, D. (2008). *Growing cooler: The evidence on urban development and climate change*. Washington, DC: Urban Land Institute.

Glaeser, E. (2011). *The Triumph of the City: How our greatest invention makes US richer, smarter, greener, healthier and happier*. Oxford: Macmillan.

Glaeser, E., & Kahn, M. (2008). *The greenness of cities: Carbon dioxide emissions and urban development working paper*. Harvard Kennedy School, Taubman Centre for State and Local Government.

Hoornweg, D., Sugar, L., & Gomez, C. L. T. (2011). Cities and greenhouse gas emissions: Moving forward. *Environment and Urbanization, 23*(1), 207–227.

ICLEI. (2008a). Cities for climate protection Australia: Local government action on climate change—measures evaluation report 2008. In *W. Department of environment, heritage and the arts*, (Ed.). Retrieved from http://archive.iclei.org/fileadmin/user_upload/documents/ANZ/Publications-Oceania/Reports/0812-CCPMeasuresReport08.pdf.

ICLEI. (2008b). Carbon neutrality framework for local governments. Retrieved from http://www.fremantle.wa.gov.au/files/223c5cc1-0c26-4b73-a4a6-9dac00ee6cad/A_ICLEI_Carbon_Neutrality_Framework_081031.pdf.

ICLEI. (2008c). *International local government GHG emissions analysis protocol (IEAP)*. Retrieved from http://archive.iclei.org/fileadmin/user_upload/documents/Global/Progams/CCP/Standards/IEAP_October2010_color.pdf.

ICLEI. (2010). Cities in a post-2012 climate policy framework. *ICLEI global reports*. Retrieved from http://ccsl.iccip.net/cities_in_a_post_2012_policy_framework.pdf.

ICLEI. (2012a). About ICLEI. Retrieved from http://www.iclei.org/iclei-global/who-is-iclei/faq.html.

ICLEI. (2012b). Cities for climate protection (CCP) campaign. Retrieved from http://archive.iclei.org/index.php?id=10829.

Kennedy, S., & Sgouridis, S. (2011). Rigorous classification and carbon accounting principles for low and Zero Carbon Cities. *Energy Policy, 39*, 5259–5268.

KPMG, C40 Cities, & CDP. (2011). *CDP Cities 2011: Global report on C40 Cities*. Retrieved from http://www.kpmg.com/IL/en/Alerts/CDP_Cities_JULY_web.pdf.

Matthews, H. S., Hendrickson, C. T., & Weber, C. L. (2008). The importance of carbon footprint estimation boundaries. *Environmental Science and Technology, 42*, 5839–5842.

Newman, P. (2014). Density the sustainability multiplier: Some myths and truths with application to perth. *Australia. Sustainability., 6*(2014), 6467–6487.

Newman, P., & Kenworthy, J. (1999). *Sustainability and cities: Overcoming automobile dependence*. Washington DC: Island Press.

Newman, P., Beatley, T., & Boyer, H. (2009). *Resilient cities: Responding to peak oil and climate change*. Washington DC: Island Press.

Newman, P., & Kenworthy, J. (2015) *The End of Automobile Dependence: How Cities are Moving Beyond Car-based Planning*. Washington DC: Island Press.

Newton, P. (Ed.). (2008). *Transitions: Pathways towards sustainable urban development in Australia*. Melbourne: CSIRO Publishing.

Ramaswami, A., Hillman, T., Janson, B., Reiner, M., & Thomas, G. (2008). A demand-centered, hybrid life-cycle methodology for city-scale greenhouse gas inventories. *Environmental Science and Technology, 42*(17), 6455–6461.

Rauland, V., & Newman, P. (2011). Decarbonising Australian cities: A new model for creating low carbon, resilient cities. Paper presented at the MODSIM2011: 19th international congress on modelling and simulation, 12–16 December 2011 Perth, WA.

Rees, W., & Wackernagle, M. (1996). Urban ecological footprints: Why cities cannot be sustainable—and why they are key to sustainability. *Environmental Impact Assessment Review, 16*, 223–248.

Roseland, M. (2012). *Toward sustainable communities: Solutions for citizens and their governments* (4th ed.). Gabriola Island, BC: New Society Publishers.

Satterthwaite, D. (2008). Cities contribution to global warming: Notes on the allocation of greenhouse gas emissions. *Environment and Urbanization, 20*, 539–549.

Sovacool, B. K., & Brown, M. (2009). Twelve metropolitan carbon footprints: A preliminary comparative global assessment. *Energy Policy, 38*(9), 4856–4869.

Sugar, L., Hoornweg, D. & Kennedy, C. (n.d.). Draft Working paper—greenhouse gas emissions from cities: Comparison of international inventory frameworks. In UNEP, UN-Habitat, & W. Bank (Eds.). Retrieved from http://siteresources.worldbank.org/INTUWM/Resources/340232-1303927718626/GHG_ReportingDraftAnnex_1.pdf.

The Bonn Center for Local Climate Action and Reporting. (2012). *Carbonn cities climate registry*. Retrieved from http://citiesclimateregistry.org/home/.

Wiedmann, T., & Minx, J. (2008). A definition of "Carbon Footprint". *Ecological Economic Research Trends 1*, 1–11 (NY, USA: Nova Science Publishers).

Wolman, A. (1965). The metabolism of cities. *Scientific American, 213*(3), 179–190.

World Bank. (2010). *Cities and climate change: An urgent agenda* (Vol. 10). Washington DC: Urban Development and Local Government.

Xi, F., Geng, Y., Chen, X., Zhang, Y., Wang, X., Xue, B., & Zhu, Q. (2011). Contributing to local policy making on GHG emission reduction through inventorying and attribution: A case study of Shenyang, China. *Energy Policy, 39*, 5999–6010.

Chapter 9
Rating Carbon in Urban Development

9.1 Precinct-Scale Accounting Schemes

Cities are predominantly designed and constructed around precincts and neighbourhoods (Sharifi and Murayama 2013). While the importance of this smaller scale (compared to the wider city level) in achieving sustainability outcomes has long been recognised (Brugmann 2010; Choghill 2008; Haapio 2012; Roseland 2005), it is only in recent years that assessment tools have been developed to specifically focus on this level (Sharifi and Murayama 2013).

Prior to this neighbourhood focus, assessment tools had predominantly targeted individual buildings as a way to assess the built environment (Retzlaff 2008), evidenced by the many building assessment tools currently in existence worldwide. Abundant research has described, evaluated and compared the vast range of such tools and analysed their role in promoting better performance (e.g. Burnett 2007; Cole 1999, 2005, 2006; Ding 2008; Dixit et al. 2012; Forsberg and Malmborg 2004; Haapio and Viitaniemi 2008; Lutzkendorf and Lorenz 2006; Ng et al. 2012; Retzlaff 2008).

Despite energy efficiency and GHG emissions being a popular focus of these tools over the years (ARUP Sustainability 2004), few in reality are able to quantitatively measure emissions (Beattie et al. 2012; Newton et al. 2013; Ng et al. 2012). Varying considerably in their design and method, these tools have adopted different 'typologies' for their assessment (Lutzkendorf and Lorenz 2006). Such typologies include checklists, life cycle assessment (LCA), rating tools, assessment frameworks, labelling and certification schemes (Haapio 2012). As neighbourhood and precinct assessment tools have largely evolved from building-scale performance tools, these new tools also vary greatly in design and measurement.

While some argue that the availability of precinct assessment tools is still rather limited (Kyrkou and Karthaus 2011), others suggest that the market has already become beset by tools proliferating in this space (AILA 2010; Blundell 2010). The lack of coordination and integration between the existing tools has also been

V. Rauland and P. Newman, *Decarbonising Cities*, Green Energy and Technology, DOI 10.1007/978-3-319-15506-7_9

recognised, adding to confusion within the industry (AILA 2010; Burke and Brown 2006).

Several studies have identified and compared the growing number of neighbourhood and precinct tools and schemes (see AILA 2010; Beattie et al. 2012; Haapio 2012; Kyrkou and Karthaus 2011; Newton et al. 2013; Sharifi and Murayama 2013). However, as Haapio (2012) rightly notes, comprehensive scientific and/or academic analyses at the precinct scale are still somewhat limited due to the relative infancy of the precinct-scale tools.

The tools and schemes that identified use a variety of terms to describe this scale of urban activity including 'neighbourhood', 'community', 'settlement', 'local area' and 'precinct'. These terms are, therefore, used interchangeably within this section. Table 9.1 lists over 60 Australian and international sustainability and climate change-related precinct-scale assessment tools that have been identified in these studies.

As mentioned, despite the increasing prevalence of precinct-scale assessment tools, very few provide adequate or comprehensive methodologies or frameworks for calculating the GHG emissions at this level (AILA 2010; Beattie et al. 2012). This lack of quantitative GHG measurement at both the building and precinct level is most likely due to the fact that the majority of the tools have focussed on the design stage and have overwhelmingly consisted of checklists embedded into point-based systems to create a sustainability rating. AILA (2010) identified this absence of quantifiable performance-based numbers within their analysis report '*Climate Adaptation Tools for Sustainable Settlements*', noting that 'the general lack of robust indices for measuring urban environmental quality or "Green Infrastructure" performance' (8) was a key concern.

Two Australian tools were identified by Beattie et al. (2012) that can provide adequate quantitative carbon measurement: PRECINX and eTool. However, even these do not provide a comprehensive analysis of all of the key areas identified in Chaps. 3 and 4 as being critical in reducing carbon at the precinct level. Nevertheless, these two tools were chosen for comparison along with six other precinct-scale tools identified as being leaders in the area of precinct-scale sustainability assessment.

All schemes discussed in this section are voluntary schemes. The eight tools chosen for analysis are briefly introduced below, which is followed by a discussion, which identifies their strengths, weaknesses and their ability to undertake a comprehensive GHG analysis for precincts. A table comparing the nine tools is provided in Appendix B.

Table 9.1 Precinct-scale climate change and sustainability assessment tools

	Name of tool	Developer and location
	Australian tools—existing and under development	
1.	AGIC Sustainability Rating Tool for Infrastructure	Australian Green Infrastructure Council
2.	C^{CAP} Precinct	Kinesis
3.	East Lake Planning Tool—Integrated Sustainability Assessment Platform	CSIRO and ACT Planning and Land Authority, ACT
4.	Ecological Footprint Model	South Australian Land Management Corporation
5.	EnviroDevelopment	Urban Development Institute of Australia, QLD.
6.	EPRA Sustainability Assessment Tool	East Perth Redevelopment Authority & GHD, WA
7.	GBCA Green Star Communities Framework tool	Green Building Council of Australia & VicURBAN, VIC
8.	HIA Greensmart	Housing Industry Association
9.	Integrated Model for Urban Sustainability	University of South Australia
10.	LESS Local Area Envisioning & Sustainability Scoring system	HASSELL
11.	Liveable Neighbourhoods	Department of Planning & Infrastructure, WA
12.	Local Government Climate Change Adaptation Toolkit	ICLEI Oceania
13.	PRECINX	Landcom, NSW
14.	Smart Growth Assessment Tool	City of Waneroo, WA
15.	STEPS Sustainable Tools for Environmental Performance Strategy	Moreland City Council, VIC
16.	Sustainability Management System	Department of Planning & Infrastructure WA
17.	Sustainability Modelling Framework SMF	South Australian Land Management Corporation
18.	Sustainable Community Rating Tools	VicURBAN
19.	Urban IT model	City Futures Research Centre
	International tools—existing and under development	
20.	Adaptation Wizard	UK Climate Impacts Programme
21.	ASPIRE	Arup International & Engineers against Poverty
22.	BCA Green Mark for Parks	Singapore Building & Construction Authority
23.	BioCity Health Index	McGregor Coxall, NSW
24.	BREEAM	BRE Global UK
25.	BREEAM Communities	BRE Global UK
26.	Bristol Development Framework	Bristol City Council
27.	CASBEE-UD	Japan Sustainable Building Council

(continued)

Table 9.1 (continued)

	Name of tool	Developer and location
28.	CASBEE-UD+	Japan Sustainable Building Council
29.	CRISP Framework	EC Thematic Network
30.	DPSIR	United Nations Environment Program
31.	Ecocity	EU Research Project
32.	EcoDistrict Performance and Assessment Toolkit	Portland Sustainability Institute (POSI), USA
33.	European Urban Audit	EU Initiative
34.	Green Mark for Districts	Singapore Building and Construction Authority (BCA)
35.	Green Neighborhood Index	Malaysian Institute of Architects (PAM) and the Association of Consulting Engineers Malaysia (ACEM)
36.	Green Plan	City of Guelph
37.	GreenPrint	BRE, UK
38.	IRM Integrated Resources Management tool	ARUP International
39.	LEED-ND	Green Building Council, USA
40.	Local Climate Change Visioning Project	Centre for Advanced Landscape Planning, Vancouver, Canada
41	LUSH Programme Landscaping for Urban Projects & High Schools	Singapore Building & Construction Authority
42.	Manchester Guide to Development	Manchester City Council, UK
43.	Neighbourhood Sustainability Framework	Beacon Pathway, NZ
44.	One Planet Living (OPL)	BioRegional Development Group and WWF International, UK
45.	PLACE3S	California, Oregon & Washington State Energy Commissions, USA
46.	PlanSmart NJ	New Jersey, USA
47.	PPDS Precinct Planning & Design Standard	Green Globe, Earthcheck
48.	QATAR Sustainability Assessment System (QSAS) Neighborhoods	Gulf organization for Research and Development, Qatar
49.	SEAT Subdivision Energy Analysis Tool	California Energy Commission PIER Programme
50.	Singapore Index on Cities' Biodiversity CBI	Singapore National Parks Board
51.	SPARTACUS System for Planning in Towns & Cities for Urban Sustainability	European Commission consortium
52.	SPeAR Sustainable Project Appraisal Routine	ARUP International

(continued)

Table 9.1 (continued)

	Name of tool	Developer and location
53.	SSIM Sustainable Systems Integrated Model	EDAW AECOM
54.	STAR Community Index	ICLEI-Local Governments for Sustainability USA, U.S. Green Building Council & the Centre for American Progress
55.	SUE-MoT	Dundee, Glasgow, Caledonian, Loughborough & St. Andrews Universities consortium
56.	Sustainability Planning Toolkit	ICLEI USA
57.	Sustainable City Program of Vancouver	City of Vancouver, Canada
58.	Sustainable Sites Initiative	Green Building Council & Partners, USA
59.	Sustainable Urban Landscapes—Site Design Manual for BC Communities	University of British Columbia
60.	TUSC Tools for Urban Sustainability Code of Practice	Synergine & Waitakere City Council, NZ
61.	UNEP Yearbook	United Nations Environment Programme

9.2 International Tools

9.2.1 BREEAM Communities

Building research establishment environmental assessment method (BREEAM) was developed in the UK in 1990 by BRE Global. It was the first commercially available environmental assessment tool for buildings and remains one of the most widely used tools for this purpose globally (Haapio and Viitaniemi 2008; Sharifi and Murayama 2013). In 2008, BRE Global created the BREEAM Communities tool to assess the wider impacts of our built environment.

The communities tool broadly follows the same methodology and point-based system as its buildings rating scheme, assessing environmental, social and economic impacts of developments. It awards between 1 and 3 credits per criteria, depending on performance. There are 51 criteria split between eight categories (Haapio 2012). Although all criteria are given equal points (i.e. up to 3 credits), not all the categories and credits are mandatory under the BREEAM tool; they are determined based on various factors such as size, type and whether projects are regional or national (Kyrkou and Karthaus 2011). Thus, the overall score for BREEAM Communities is given as a percentage, which is translated into six ratings such as ‘pass’ (≥25 %), ‘good’ (≥40 %) and ‘excellent’ (≥70 %) (Haapio 2012).

9.2.2 LEED Neighbourhood Development

Leadership in Energy and Environmental Design (LEED) was developed by the US Green Building Council and the Natural Resources Defence Council in 1994, taking inspiration from the UK BREEAM rating scheme. While LEED predominantly targets the US market, it is also used informally around the world. Together, LEED and BREEAM are the two globally dominant building environmental assessment tools (Kyrkou and Karthaus 2011).

LEED for Neighbourhood Development (LEED-ND) was initially developed and piloted in 2007 (Sharifi and Murayama 2013), but the official rating system was only released in 2010 (Haapio 2012). Like BREEAM Communities, LEED-ND uses a points-based system to determine the overall rating, but the number of credits per category is different in the two tools. There are 53 criteria in the LEED-ND tool, spread over three primary categories and two additional ones. Unlike BREEAM, the number of credits per criteria varies under LEED (e.g. some criteria are awarded 10 credits) and all criteria are mandatory (Haapio 2012; Kyrkou and Karthaus 2011).

Since LEED-ND was developed in conjunction with the Congress for New Urbanism, LEED-ND tends to place greater emphasis on rewarding principles of new urbanism such as those relating to land use and transport, than the broader environmental concerns and impacts that BREEAM focuses on (Kyrkou and Karthaus 2011). There are also critiques of new urbanism that show the outcomes of their principles are insufficient if location, transit provision and density are not more seriously considered in designs (Falconer et al. 2010). As the final scores are just points, it is easy to miss significant parameters that can change real carbon outcomes much more than would a list of ‘worthwhile’ measures. The overall score for LEED-ND is provided in points and given one of four ratings: certified, silver, gold or platinum.

9.2.3 CASBEE Urban Development

Developed in 2004, Japan’s Comprehensive Assessment System for Building Environmental Efficiency (CASBEE) was a joint initiative of the government, industry and academia (Haapio 2012). Two years later, in 2006, CASBEE for Urban Development (UD) was created and was thus the first of the major precinct-wide sustainability assessment schemes available. It also uses a points system to evaluate developments based on categories with specific criteria.

CASBEE-UD has 80 criteria that, similar to BREEAM, are all equal. Each criterion is awarded between 1 and 5 points or credits, depending on performance, though in some cases, the full range of points is not applicable (Sharifi and Murayama 2013). Unlike BREEAM and LEED, the way in which CASBEE calculates its overall score is based on qualitative and quantitative measurement and ratios and thus may be considered quite complicated in comparison with the above tools.

The final score or rating in CASBEE-UD is based on a points system in which <0.5 = 'poor', 0.5–1.0 = 'fairly poor', up to 1.5–3.0 = 'very good' and ≥3.0 'excellent' (Haapio 2012).

9.2.4 One Planet Communities

One Planet Communities is an international design and sustainability-branding tool for community-scale development. It was created in 2004 by BioRegional, a not-for-profit organisation based in the UK, who now has offices in eight countries around the world.

One Planet Communities differs from other rating schemes in that it does not provide a sliding scale for achievements (e.g. 4, 5 or 6 stars); a developer either meets the requirements or they do not. BioRegional also does not provide certification as such, but will 'endorse' communities that demonstrate their commitment to achieving the One Planet Living principles. The concept of One Planet Living is based on the ecological footprint model (Rees and Wackernagle 1996), which demonstrated that we are currently living beyond what the planet can provide in one year.

The tool identifies 10 common international targets (CITs) or themes that aim to improve quality of life of citizens and reduce per capita consumption and pollution at the community scale to a level that allows us all to live within our planetary limits. Stringent targets are set for carbon, waste, transport, materials and water and include measurable reductions such as 'zero carbon' (e.g. for the operation of buildings) and 95 % reduction in waste, though if and how GHG reductions are measured is not clear.

To receive endorsement, communities must demonstrate that they have set an action plan, are delivering on the action plan and are willing to undertake an annual review process. They don't necessarily need to have achieved 'Zero-Carbon'.

9.3 Australian Tools

9.3.1 Green Star Communities

Australia's Green Star Communities tool is the nation's leading precinct-scale sustainability assessment tool, developed by the Green Building Council of Australia (GBCA). The GBCA was first established in 2003 and is a member of the World Green Building Council (Mitchell 2009). Although initial development of the communities rating tool began in 2009, the first pilot version was only released in 2012 and is therefore the newest of these larger, well-known schemes.

Table 9.2 Rating framework adopted by the Green Building Council of Australia

Point score	Green star rating	Outcome
45–59	4 star	Best practice
60–74	5 star	Australian excellence
75+	6 star	World leader

While being influenced by BREEAM Communities and LEED-ND, Green Star Communities undertook extensive consultation with various stakeholders from industry, academia, government, planning, utilities and social service organisations in order to create a unique Australian tool (Blundell 2010). Similar to the aforementioned tools, Green Star Communities uses a point system to rate a development's overall sustainability performance. The tool has 38 criteria or credits spread over six main categories. A varying number of points are available under each credit (i.e. from 0.5 to 10) and together equal a total score of 110 (GBCA 2013). There are some mandatory requirements within the tool (e.g. yes/no checklists) and all six categories are mandatory. Their are an additional 10 points, which can be gained which are voluntary and achieved through demonstration of innovation. The overall rating is given in points (45–59, 60–74, 75+) and stars (4–6) and expressed in terms of comparative practice, e.g. 'best practice', 'Australian Excellence' or 'World Leader' (see Table 9.2).

9.3.2 EnviroDevelopment

EnviroDevelopment is another well-known and well-utilised Australian neighbourhood assessment tool, particularly popular with smaller players within the development industry. It was created by the Urban Development Institute of Australia (UDIA) in 2006, and while it has the ability to assess various types of developments including commercial, residential and industrial, it has predominantly focused on low-density residential development (EnviroDevelopment 2011).

The tool has six categories or 'elements', which are represented by leaves in its rating system. Unlike the previous tools, it does not use a points-based system to rate design, but has specific requirements that need to be met under each criteria. Some elements (e.g. energy) also set performance targets such as a percentage reduction in GHGs on a state's average for energy generation. Supporting evidence needs to be provided (UDIA n.d.), though it can be as basic as documented evidence from a technology supplier that their technology has lower emissions than prevailing technologies. Thus, it eliminates the need for comprehensive carbon accounting.

Despite the quite prescriptive and detailed nature of the technical standards, which outline how to achieve the criteria under each element, the elements themselves are not mandatory (AILA 2010). Developers can, therefore, choose one or more elements to have certified (i.e. a leaf). However, if fewer than six elements are sought, eight additional (non quantitative) criteria must also be met (UDIA n.d.).

9.3.3 PRECINX

PRECINX is an Australian neighbourhood sustainability assessment tool, developed by Kinesis for the New South Wales State Land Development Agency, Landcom. Launched in 2009, it was one of the first of a new generation of precinct-scale assessment tools that use a more quantitative approach for assessing sustainability. Rather than using points to rate a design, it helps to inform the design by modelling various options around onsite energy, embodied CO_2 in materials, potable water, storm water, transport and housing diversity.

As the tool is able to quantify the differences in tonnes of CO_2, litres of water, vehicle kilometres travelled and housing affordability using various modelling scenarios and options, it helps developers make better urban planning decisions based on real numbers (Perinotto 2009). As suggested in the previous chapter, assessment should incorporate both life cycle and operational aspects of an urban development. PRECINX is able to combine these (Matan et al. 2015).

The tool has been increasingly adopted by major developers across Australia (such as Australand, LendLease and Stockland) and is being used in all states by the government land development agencies under a licence. Western Australia's State land development agency, Landcorp, has begun to use it on some projects (Perinotto 2011). The tool was developed for the same NSW government agency that developed the building-scale assessment tool BASIX. The BASIX tool has become an acceptable part of planning, but PRECINX, although planned to be used in regulatory approval processes, has not yet been accepted for that purpose. If the tool continues to grow, it is likely to be used as a design and assessment tool. Other tools that are more transparent and publically available may also be developed for this purpose as the need to quantify design outcomes for decarbonising cities becomes more and more pressing.

9.3.4 eTool

eTool emerged around the same time as PRECINX, in 2009, and began as a life cycle carbon assessment-modelling tool to help industry understand the importance of embodied carbon in the construction of individual buildings. The aim was to provide a user-friendly online platform that could assist architects, designers and developers to make informed decisions about how best to reduce carbon within buildings from a life cycle perspective.

While individual buildings were the initial focus of the tool, it now has the capacity to assess various types of built forms, from small residential and/or commercial developments to infrastructure projects, to larger scale urban redevelopment projects. Being a life cycle tool, it models the carbon associated with the design, construction and operational energy of developments. However, it currently does not measure the carbon associated with the transport of residents or the emissions associated with the operation of precinct-scale infrastructure (e.g. water and waste management).

Unlike the other international precinct assessment tools, which focus on a variety of environmental and sustainability issues, eTool focuses primarily on carbon. This tool and PRECINX are the only two that apply rigorous carbon accounting methodologies.

9.4 Issues Around Precinct-Scale Tools

The importance of assessment and rating tools for the built environment has long been identified and discussed (Burke and Brown 2006; Cole 1999, 2005, 2006; Ding 2008; Forsberg and Malmborg 2004). They provide a structured way of collating, presenting and disseminating information on the environmental performance of buildings and have thus played a critical role in helping to shift behaviour while also providing the ability to measure and track our progress towards sustainability (Ding 2008). Burke and Brown (2006) note the importance of rating schemes in providing benchmarking as a way to stimulate environmental performance. For these and other reasons, rating and assessment tools have been identified as 'one of the most potent and effective means to both improve the performance … and to promote higher expectations and demand' within the building industry (Cole 2006: 357). Nevertheless, there are several issues with how the various tools measure and evaluate. These are discussed below.

9.4.1 Measurement Methods

The brief analysis of the eight precinct tools provided above has demonstrated how greatly assessment tools can vary in terms of design and methodology, and thus, their overall output and function as a tool. Cole (2005) makes the distinction between assessment and rating tools, suggesting that assessment tools are more quantitative and often based around LCA, while rating tools provide the 'extended outputs from an assessment process' (456) as a way to compare and brand achievements. The outputs generated from the assessment processes in rating schemes often use diverse criteria, both qualitative and quantitative in nature, commonly comprising of checklists and multi-criteria analyses (Retzlaff 2008). They therefore often need to be embedded into point-based frameworks to provide a common metric to calculate the overall rating. Forsberg and Malmborg (2004) also acknowledge a broad division of tools, classifying them either as quantitative (e.g. LCA focussed) or qualitative (based on mixed criteria) tools.

Sharifi and Murayama (2013), who reviewed seven precinct sustainability assessment tools, note the use of layers of measurement within the tools, such as themes, criterion and indicators. It is common that a series of indicators will be used to measure specific criteria (or credits), which when combined, make up a broad theme as illustrated in Fig. 9.1.

Fig. 9.1 Example of layers of measurements within assessment tools

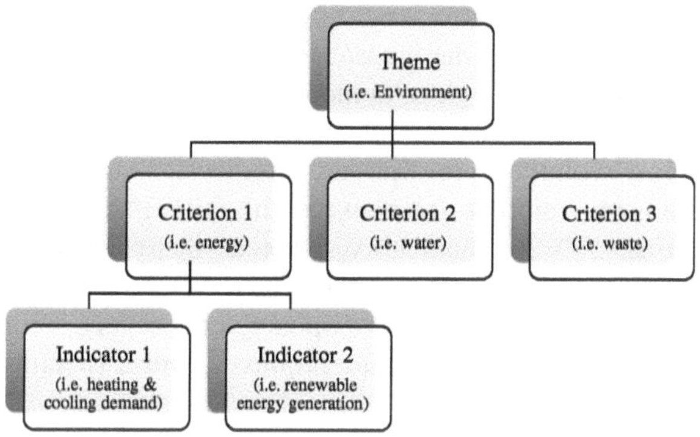

The overall score achieved is then usually displayed as a designation or rating such as 'good', 'excellent' or 'silver', 'gold' (Cole 2005) or in stars, such as the Green Star Communities tool, or other, often abstract graphical representations such as leaves as in the case of EnviroDevelopment.

9.4.2 Point-Based Systems and Use of Weightings

The majority of precinct assessment tools and schemes discussed in the section above adopt a point-based system for measurement. While several of the tools reviewed within this analysis, as well as tools reviewed in previous studies, have similar themes or criteria, the number of points or credits given to each area often varies considerably between tools (Kyrkou and Karthaus 2011). The idea of allocating varying amounts of points to certain areas is also known as 'weighting'. Weighting involves assigning different values to different criteria in order to reflect their perceived importance.

As Ding (2008) recognises, however, there is often 'no clear logical or common basis for the way in which the maximum number of points is awarded to each criterion' (458), as tools rarely provide the reason or justification for their allocation. Sharifi and Murayama (2013) contend, 'weighting is one of the most theoretically controversial aspects within SA [sustainability assessment]' (80), noting that the process of selecting criteria and assigning the weights is usually highly subjective. Weightings are usually established using stakeholder surveys to determine the overall importance of various environmental issues (Burnett 2007). This suggests that the developers of tools and their advisory teams can preference, by giving greater weight to, an area that is desirable to them.

AILA (2010) also identifies the notion of 'vested interests' as a concern in the development of tools. Blundell (2010) further highlights the risk associated with tools developed by certain industries such as the property industry, suggesting they can potentially push specific agendas such as promoting urban growth. Regardless of the reasons behind the choice of weightings, their adoption can effectively distort numbers, leading to very different rating scores between different tools measuring essentially the same areas/categories. This makes comparisons between tools, and consequently the developments they are rating, difficult.

Adding further confusion to the overall output of numbers is the decision to allow voluntary or optional credits for assessment. For example, despite the numerous similarities between the BREEAM and LEED tools, all criteria within LEED-ND are mandatory, whereas there are several optional criteria under the BREEAM Communities tool (Kyrkou and Karthaus 2011). The EnviroDevelopment tool also allows developers to choose any number of 'leaves' for certification, meaning developers are not required to meet specific environmental standards in all areas.

9.4.3 Complexity

The major appeal of point-based systems is their simplicity. Any developer can go through the motions of an assessment and claim the points at a certain level of performance. This may make people in the development industry, in communities or in government feel better, and indeed may lead to innovations that would not have happened under BAU. However, they also may be misleading as the assessed outcomes are not necessarily quantified. The urban systems being built are generally much more complex than a list of isolated parameters can ever suggest.

Another aspect of the point-based assessment tools is that they themselves are growing into quite complicated structures in order to try and accommodate the complexities of urban systems. Kyrkou and Karthaus (2011) suggest that neighbourhood sustainability assessment schemes are becoming increasingly complex and detailed, particularly regarding the allocation of weightings, which they suggest decrease functionability, comparability and adaptability of tools and add to the ambiguity of results. This was also found to be the case in the review of the eight tools above. Kyrkou and Karthaus (2011) thus recommend that future frameworks would benefit from less complexity and a more 'transparent and auditable structure' (210) in order to ensure greater understanding and wider adoption of tools. This supports the notion and need for a common, universally adopted framework for classifying precinct-scale emissions and simple quantifiable models that can be used in assessment.

9.4.4 Phase of Adoption of the Tool

One of the major limitations or shortcomings of assessment rating tools is that they are often adopted too late in the design process, meaning the decisions that can potentially maximise the overall sustainability and environmental gains are not built in from the start (AILA 2010). Burke and Brown (2006) contend that rating tools are often developed as appraisal tools, which necessarily need to be applied at the end of the building phase, meaning that many of the 'opportunities to improve performance are often lost long before the rating tool is even applied' (5–6).

As a result, sustainability measures are in many cases 'bolted on' at the end of the design process as developers seek to gain credits from tick box items, which can result in perverse outcomes. There are numerous examples of this occurring within industry. In Australia, tri-generation systems appear to be one of the common 'bolt-on' options for buildings attempting to attain additional Green Star points. However, due to speedy installation processes and a lack of communication between building designers and engineers, many systems are incorrectly sized in relation to a building's heating, cooling and energy requirements and are thus never turned on or utilised (Kelly 2012). This causes many to doubt the usefulness and efficiency of the systems.

This 'point-grabbing' phenomenon fostered by some rating tools can therefore have a detrimental effect on the overall sustainability agenda, as it can lead to some promising technologies, such as tri-generation, being perceived negatively by industry and the public.

9.4.5 Performance and Evaluation

The lack of follow-up or performance-based monitoring or assessment within current rating tools is also seen to be a limitation of existing tools. AILA (2010) identifies this, observing that current precinct-scale tools appear to concentrate either on predictive performance (i.e. design-based) or on operational performance and noting that few are able to measure both in a comprehensive manner. This is true for the majority of the tools discussed in this section, with the exception of eTool and EnviroDevelopment. eTool has a performance measure built into the tool, while EnviroDevelopment has an option for annual certification based on performance targets. Some of the larger, national tools such as Australia's GBCA's Green Star certification system have only recently designed performance tools for buildings. GBCA's new Green Star Community Tool does include a performance component to it, as it needs to undergo continual certification every 5 years.

Providing an evaluation component to tools to encourage the measuring and assessment of performance will help to address and minimise the perverse outcomes associated with some of the 'bolt-on' actions mentioned above. It can also help to identify when problems occur as in the case of the BedZED development discussed in Chap. 6, where the malfunctioning of the biomass plant meant that the development failed to maintain its intended goal of being supplied by 100 % renewable energy.

Jones and Patterson (2007) warn of the danger of this occurring when using tools without performance elements, highlighting that 'thinking is constrained to what is only in the tool and to the time period in which the tool is applied' (257).

Fig. 9.2 Type of tool and ideal order for use

9.4.6 Functionality of Tools: Decision-Making Versus Assessment and Rating

A further area of discussion relates to the ability of rating tools to actually help inform the decision-making process or pre-design process, as opposed to simply rating developments at the end based on arbitrary checklists. The ability to weigh up various options based on real numbers and quantifiable benefits, to help planners, architects and urban designers choose the best options for meeting their prescribed goals, is something largely lacking within the existing rating and assessment tools targeting the built environment (AILA 2010; Retzlaff 2008).

From the tools analysed, eTool and PRECINX are the only two able to provide this sort of 'options-modelling' or decision-support, i.e. they are design tools. Retzlaff (2008) discusses the need for these types of tools, particularly in respect to reducing greenhouse gas emissions. Nevertheless, neither tool currently considers all the areas of a development identified in Chaps. 3 and 4. Enabling developers to choose the most appropriate option at the design phase allows important decisions to be made about various materials, technologies and urban planning options based on quantifiable factors such as GHG emissions, water use and cost.

In understanding the differences between tools, e.g. decision-making capacity versus assessment or rating, it becomes clear that each type of tool provides important but quite distinct functions. Although many tools try to combine the functions,[1] they may in fact prove to be more useful when used for the purpose they were primarily designed to perform. For example, a decision-making tool may be best used in the initial pre-design phase in order to help guide and inform the design, while an assessment tool can help to determine a development's overall success (usually in a specific area) and thus their impact based on actual performance. This information can then feed into a rating or certification scheme, which provides the overall branding and comparability function.

Sustainability tools for the built environment have thus been identified as falling under three distinct categories: decision-support tools, assessment tools and rating and certification tools, as illustrated in Fig. 9.2.

[1] For example, eTool was initially designed as a decision-making tool to help inform the design process, though due to customer demand, they also created a rating system. The rating system is obviously limited to the specific outputs of the tool, measuring predominantly GHG emissions, though it does rate sustainability in a broad sense.

Regardless of the order tools are used in, or which tools are ultimately chosen, it is important for both the developers of tools and their users to recognise and acknowledge the limitations of each type of tool. For example, a rating tool applied at the end of a project's design phase is unlikely to produce the most optimal outcome in terms of environmental performance, while current decision-making or assessment tools, which generally focus on relatively narrow areas (e.g. GHG's emissions) are unable to cover the full spectrum of criteria needed to rate the overall sustainability of a precinct (AILA 2010).

9.5 The Need for a Sustainability Framework

Many studies have acknowledged the rapid increase of tools and schemes that measure and rate various aspects of our built environment. Noting the general 'lack of knowledge and skills relating to what tools to use and how to use them' (Jones and Patterson 2007: 256), the proliferation of tools appears to be causing more confusion among users than providing assistance (Blundell 2010). Indeed, according to a study Burke and Brown (2006) conducted, participants insisted that when 'new innovative ways to measure urban sustainability' are created, opportunities should be sought 'to embed their work into existing tools, if possible' (7).

Recognising the abundance of, and variability in, functionality between tools currently available at the precinct level, AILA (2010) argue for a more integrated and coordinated overall approach to progressing the sustainability agenda for the built environment in Australia. One of the main recommendations that arose from the AILA (2010) report was the development of a national sustainability assessment framework as a way of addressing the inconsistencies associated with having such a vast number and range of tools targeting precinct-scale urban development. They assert as follows:

> Until such a mechanism is in place, climate adaptation and sustainability assessment tools will continue to be designed to serve arbitrarily defined goals, responding to uncoordinated sectoral interests, and without reference to any consistent implementation pathway for broader national objectives. (AILA 2010: 5)

The GHG framework proposed in this book thus forms one essential element of what could be considered within the broader sustainability framework for the built environment proposed by AILA (2010). Given the current lack of consistent quantitative GHG measurement within existing tools (see Appendix B for a comparison of the tools discussed), and the acknowledgement of the importance of GHG emissions in our future low-carbon economy, the creation and adoption of this universal framework is critical. The framework should be considered at, and inform, each of the three stages identified above (i.e. during decision-making, assessment and rating), as the areas identified in the framework have a direct impact on each stage.

It should be noted that it is not the intention of this book to create a tool that can calculate the GHG emissions of a development or provide decision-making support.

The overarching GHG framework proposed in this book is designed to provide guidance on the emissions that ought to be calculated within urban development at the precinct level. Many decision-support and/or GHG measurement tools may be developed using this framework. These tools do not necessarily need to provide any certification or rating themselves, but can be used in conjunction with existing rating schemes such as BREEAM, LEED, Green Star and EnviroDevelopment to demonstrate that the GHG component has been met.

9.6 Conclusion

Considerable emissions abatement can be achieved when tools for low-carbon urban development are used in the design stage as well as the assessment stage. Directing greater attention towards the precinct-scale for GHG management may help to bridge the gap between production and consumption-based approaches to calculating emissions as it allows more demand-side emissions to be considered, for example, those arising from household transport patterns, energy and water consumption, as well as embodied emissions such as those in construction materials.

However, in order to include these sorts of emissions, a different carbon accounting framework and scheme is needed to those currently targeting the city level. A precinct-scale GHG framework will require more of a life cycle approach to calculating emissions (which includes the embodied and operational emissions of buildings as well as transport by residents), rather than the city-scale inventory approach, which includes only the operational/industrial activities occurring within cities. These activity-based emissions from industries that occur are less relevant to developers, who are generally not responsible for business activities occurring within their precinct developments. However, developers do need to ensure that they make well-informed design decisions around location, density and land-use mix, as these will impact severely on operational energy use.

Despite the growing abundance of sustainability assessment and rating tools targeting the built environment and urban development, very few have focussed specifically on GHG emissions or provide a comprehensive framework for their measurement. A key challenge, therefore, is to determine how best to translate and integrate the knowledge and insight gained from the development of the GHG measurement processes at the city level into a carbon accounting framework at the precinct level, a framework that fosters carbon reduction through improved urban planning and precinct design.

Focusing on this level will help to promote better interaction between key urban stakeholders such as developers, builders, architects, urban planners, utilities, local governments and the community, thereby helping to create tangible change at the ground level using a bottom-up approach to carbon reduction.

However, there is currently no established governance framework that can help to facilitate a process involving such a myriad of stakeholders at this level or to set out the requirements of a model and assessment framework. Thus, determining a robust governance structure to deal with precincts as they attempt to decarbonise represents a major challenge.

While cities and local governments already have institutional frameworks in place to be able, at a democratic level, to sign on to the initiatives outlined in this chapter, it is far more difficult to get buy-in and collaboration from developers and other key stakeholders who are not subject to, and bound by, the same institutional requirements. It is argued, therefore, that a framework and process for calculating and addressing emissions at the precinct scale (a quantitative design tool rather than a 'tick-box' rating schemes) would benefit from linking into frameworks, schemes and initiatives already established at the local government level. This may further benefit local governments by allowing them to develop a greater understanding of their community's emissions, as well as to discuss future policies and initiatives to address them. These ideas will be pursued further in Chap. 12 to show how tools can assist both developers and local governments.

The next chapter examines the certification process for carbon neutrality and how this can help to increase the credibility of carbon claims.

References

ARUP Sustainability. (2004). *Overview of sustainability rating tools phase 2—recommendations report*. Retrieved from http://www.construction-innovation.info/images/pdfs/BCC_ARUP_Tools_Reports.pdf.

Australian Institute of Landscape Architects (AILA). (2010). *Climate adaptation tools for sustainable settlements (CATSS)*. Canberra: Australian Institute of Landscape Architects.

Beattie, C., Bunning, J., Stewart, J., Newman, P., & Anda, M. (2012). Measuring carbon for urban development planning. *The International Journal of Climate Change: Impacts and Responses, 3*(4), 35–52.

Blundell, L. (2010). A confusion of tools raises issues…and a few hackles. *The Fifth Estate*. Retrieved from http://www.thefifthestate.com.au/innovation/planning/a-confusion-of-tools-raises-issuesand-a-few-hackles/12503.

Brugmann, J. (2010). *Welcome to the urban revolution: How cities are changing the world*. New York: Bloomsbury Press.

Burke, M. I., & Brown, A. L. (2006). Planning by rating scheme: genealogy, scales of application and ways forward for the formalised rating approach to urban development approval. In *2nd State of Australian cities conference* (Vol. 30). Brisbane.

Burnett, J. (2007). City buildings-eco-labels and shades of green! *Landscape and Urban Planning, 83*, 29–38.

Choghill, C. (2008). Developing sustainable neighbourhoods. *Habitat International, 32*, 41–48.

Cole, R. (1999). Building environmental assessment methods: Clarifying intentions. *Building Research & Information, 27*(4), 230–246.

Cole, R. (2005). Building environmental assessment methods: Redefining intentions and roles. *Building Research & Information, 33*(5), 455–467.

Cole, R. (2006). Shared markets: Coexisting building environmental assessment methods. *Building Research & Information, 34*(4), 357–371.

Ding, G. (2008). Sustainable construction—The role of environmental assessment tools. *Journal of Environmental Management, 86*, 451–464.

Dixit, M. K., Fernández-Solís, J. L., Lavy, S., & Culp, C. H. (2012). Need for an embodied energy measurement protocol for buildings: A review paper. *Renewable and Sustainable Energy Reviews, 16*, 3730–3743.

EnviroDevelopment. (2011). *Envirodevelopment Projects*. Retrieved from http://www.envirodevelopment.com.au/01_cms/details.asp?ID=57.

Falconer, R., Newman, P., & Giles-Corti, B. (2010). Is practice aligned with the principles? Implementing new urbanism in Perth, Western Australia. *Transport Policy, 17*(5), 287–294.

Forsberg, A., & Malmborg, F. V. (2004). Tools for environmental assessment of the built environment. *Building and Environment, 39*, 223–228.

Green Building Council Australia (GBCA). (2013). Green Star—Communities Rating Tool. *Rating Tools*. Retrieved from http://www.gbca.org.au/green-star/green-star-communities/rating-tool/.

Haapio, A. (2012). Towards sustainable urban communities. *Environment Impact Assessment Review, 32*(1), 165–169.

Haapio, A., & Viitaniemi, P. (2008). A critical review of building environmental assessment tools. *Environmental Impact Assessment Review, 28*, 469–482.

Jones, P., & Patterson, J. (2007). The development of a practical evaluation tool for urban sustainability. *Indoor and Built Environment, 16*(3), 255–272.

Kelly, D. (2012). Simons wins cogen contract for Rydges. Retrieved from http://www.thefifthestate.com.au/archives/35825/.

Kyrkou, D., & Karthaus, R. (2011). Urban sustainability standards: Predetermined checklists or adaptable frameworks. *Procedia Engineering, 21*, 204–211.

Lutzkendorf, T., & Lorenz, D. P. (2006). Using an integrated performance approach in building assessment tools. *Building Research & Information, 34*(4), 334–356.

Matan, A., Newman, P., Trubka, R., Beattie, C., & Selvey, L. (2015). *Health, Transport and Urban Planning: Quantifying the Links between Urban Assessment Models and Human Health.* Urban Policy and Research. doi: 10.1080/08111146.2014.990626.

Mitchell, L. (2009). Green star and NABERS: Learning from the Australian experience with green building rating tools. Paper presented at the 5th urban research symposium in Cities and Climate Change: Responding to an Urgent Agenda, Marseille, France.

Newton, P., Marchant, D., Mitchell, J., Plume, J., Seo, S., & Roggema, R. (2013). Performance assessment of urban precinct design—a scoping study. Retrieved from http://www.lowcarbonlivingcrc.com.au/sites/all/files/publications_file_attachments/rp2001_-_performance_assessment_urban_precinct_design-final_0.pdf.

Ng, S. T., Chen, Y., & Wong, J. M. W. (2012). Variability of building environmental assessment tools on evaluating carbon emissions. *Environmental Impact Assessment Review, 38*, 131–141.

Perinotto, T. (2009). Precinx—Landcom's new neighbourhood sustainability tool that cuts through the myths. *The Fifth Estate*. Retrieved from http://www.thefifthestate.com.au/archives/8107/.

Perinotto, T. (2011, December 13). Landcom's PRECINX notches up LandCorp as first licensee. *The Fifth Estate*. Retrieved from http://www.thefifthestate.com.au/archives/30548/.

Rees, W., & Wackernagle, M. (1996). Urban ecological footprints: Why cities cannot be sustainable—and why they are key to sustainability. *Environmental Impact Assessment Review, 16*, 223–248.

Retzlaff, R. (2008). Green building assessment systems. *Journal of American Planning Association, 74*(4), 505–519.

Roseland, M. (2005). *Toward sustainable communities: Resources for citizens and their governments*. Canada: New Society Publishers.

Sharifi, A., & Murayama, A. (2013). A critical review of seven selected neighborhood sustainability assessment tools. *Environmental Impact Assessment Review, 38*, 73–87.

Urban Development Institute of Australia (UDIA). (n.d.). EnviroDevelopment Technical Standards National Version 1.0: Urban Development Institute of Australia.

Chapter 10
Certifying for Carbon Neutrality

10.1 Introduction

The number of businesses and organisations that making environmental claims and promoting green credentials in order to differentiate themselves has increased significantly in recent years. This has correlated with consumers' willingness to pay a premium on 'green' goods and services in an increasingly carbon-constrained and environmentally conscious marketplace (Brecard et al. 2009; OECD 2002).

For many organisations, the notion of moving 'beyond compliance' to achieve higher environmental standards is appealing, particularly if it warrants a higher price on their products or services or can help to differentiate them from their competitors. However, communicating environmental attributes to consumers is often hard as environmental actions or processes are not always observable or tangible (Brecard et al. 2009; Hamilton and Zilberman 2006). Eco-labeling and/or certification has thus emerged as a way of validating environmental claims to drive more sustainable consumption practices (Jamalpuria 2012).

Since this eco-movement emerged largely in the voluntary space, it initially lacked the stringent rules and regulations that generally underpin compliance or mandatory practices, industries and markets.[1] Consequently, the rapid rate at which the unregulated, voluntary 'green industry' developed saw growing concern over the credibility and legitimacy of some of the environmental, and in particular, carbon claims being made. Consumers began to become wary of 'green' claims, particularly when carbon offsets, or the term 'carbon neutrality' were used, as sufficient information was rarely provided to substantiate or explain the claims.

This chapter examines some of the reasons why certification for carbon-related achievements is necessary, highlighting issues pertaining to the regulatory system and using the Australian Competition and Consumer Commission (ACCC) as a case study. An overview of the organisational structures and key documents

[1] The voluntary market has evolved considerably since it began and is now is relatively heavily regulated.

V. Rauland and P. Newman, *Decarbonising Cities*, Green Energy and Technology, DOI 10.1007/978-3-319-15506-7_10

underpinning the industry is then provided, followed by a brief summary of several well-known existing Australian and international carbon certification schemes. Australia's National Carbon Offset Standard Carbon Neutral Program (NCOS-CNP) is identified as one of the leading independent certification schemes worldwide, and it is thus discussed in greater depth in Chap. 11.

10.2 Avoiding False and Misleading Claims

In response to the growing issue of the credibility of carbon claims in Australia in the mid 2000's, the ACCC began publicly investigating a number of businesses for false claims. One notable case involving General Motors was taken to court by the ACCC.[2] General Motors was found guilty of breaching the *Trade Practices Act 1974* with its carbon claim being declared false and misleading by the court. The case, which was given significant press coverage, was used to demonstrate to other organisations and businesses that false and misleading claims that breach the Act will be actively pursued by the ACCC.

Several further investigations followed,[3] before the ACCC in 2008, released a document titled *Carbon Claims and the Trade Practices Act*. The purpose of this document was to inform and provide guidance to businesses and organisations on how to avoid making false claims. The document warns that those making claims found to be breaching the Act, i.e. those considered to be misleading consumers (intentionally or not) and thereby engaging in deceptive conduct, are liable for fines up to $1.1 million (Australia Competition and Consumer Commission 2008).

The ACCC urges organisations to clearly explain all claims, noting that the absence of universal carbon definitions can lead to a significant variation in consumers' understanding of the terms. They assert that 'vague, unsubstantiated, confusing or misleading information will reduce consumer confidence in carbon claims, thereby disadvantaging ethical traders' (Australia Competition and Consumer Commission 2008) and ultimately damaging the emerging green economy.

In 2011, the above guidelines were revised and new information added about the introduction of the National Carbon Offset Standard (NCOS), the Carbon Neutral Program and the Australian Carbon Trust (which became Low Carbon Australia, before merging into the Clean Energy Finance Corporation). This guide continues to play an important role in driving greater investment in low-carbon technologies and greener practices and increasing demand for eco-labeling, certification and the marketing of green credentials.

[2] See 'Saab "Grrrrreen" claims declared misleading by Federal Court' (available at: http://www.accc.gov.au/content/index.phtml/itemId/843395).

[3] See 'When green wash won't wash: Avoiding misleading environmental claims' (available at: http://www.edo.org.au/conference/2010/2010michael_terceiro.pdf).

With the introduction of the Carbon Price Mechanism in 2012, carbon reduction activity began to focus on the Front-End of the economy where the pricing applied. However, the removal of the mandatory carbon price by the Abbot Government 2014, and the introduction of the voluntary Direct Action Plan, has meant greater emphasis is again on the End-User part of the economy. Funding is being provided for the organisations or projects that can demonstrate emissions reductions, though only if sufficient aggregation of emissions reductions can be demonstrated.

Ensuring that carbon claims are clear and transparent is critical to avoiding the legal risks and liabilities associated with false claims under the Trade Practices Act. It can also help to bring greater credibility and integrity to carbon claims. However, having carbon claims recognised or certified by a third party provides even greater assurance to consumers and the ACCC, providing that the third party is a reputable and, ideally, accredited body. Such certification should ensure that carbon audits/footprints are undertaken using internationally accepted best practice carbon/GHG accounting standards and are verified by an independent third party. This, together with assurance that purchased offsets (if certification is for carbon neutrality) are legitimate and trustworthy, can add considerable validity to claims.

10.3 What Is Certification and Who Conducts It?

Any organisation or body can certify a product/service or business as being carbon neutral in the same way anyone can make a carbon claim. However, the degree of reliability and trustworthiness of a certification scheme, just like any claim, can vary greatly. As Murray and Dey (2008) pointed out in their study of eleven online carbon calculators, little similarity existed between each calculator, creating confusion and doubt around the correct methodology, the reliability and the integrity of offsets being sold and, thus, the overall legitimacy of certification being offered.

Fortunately, there are an increasing number of organisations and schemes emerging, whose certification is based on international standards and methodologies guaranteeing rigour and robustness in calculations and stringency in terms of verification. Such organisations can also seek accreditation themselves in order to be recognised and acknowledged as providing reputable certification to clients.

Although the terms accreditation and certification are frequently used interchangeably, they have quite different roles, functions and meanings when discussed in terms of providers. The International Organisation for Standardisation (ISO), an independent body that develops and publishes international standards used by governments and private industries worldwide,[4] has provided a distinction between the two terms. They identify 'accreditation' as that given to bodies or organisations who are recognised as being reputable providers of conformity assessments and certification (ISO 2013). These bodies then award certification to final consumers,

[4] See http://www.iso.org/iso/about.htm.

businesses and products that have met specific standards. Further clarification of the terms and the role of the ISO are provided in Appendix C.

10.4 Certification Schemes

As mentioned above, certification is 'third-party attestation related to products, processes, systems or persons' (ISO 2004). While hundreds of carbon-neutral certification schemes are likely to exist around the world, the following section outlines some of the more well-known international, national and private schemes offering carbon-neutral certification, with a focus on Australia. Interestingly, few schemes or organisations currently pursue accreditation from an official accreditation body such as those outlined in Appendix C, despite the credibility it can bring to the company providing the certification process. This may have to do with the fact that there are relatively few carbon-neutral certification schemes compared to the rapidly increasing number of organisations claiming carbon neutrality, thereby placing less importance on the need for certification companies to differentiate themselves. Nevertheless, most certification bodies and organisations conduct their certification and GHG measurement process based on the international GHG accounting standards mentioned in Appendix C.

A summary of seven of the most well-known carbon certification schemes is discussed below, and a comparison between them is provided in Appendix D.

10.4.1 CarboNZero

Formed in 2001 by New Zealand's Government owned institute *Landcare Research New Zealand Ltd*, carboNZero was one of the first organisations to offer carbon-neutral certification worldwide. It was also the first company accredited through JAS-ANZ under ISO 14065. Based in New Zealand but operating internationally, carboNZero has certified over 200 companies, products, events and services under its program. Countless more have been certified under another scheme (CEMARS), which does not require offsetting (i.e. to become carbon neutral), but ensures that a company is taking action to manage and mitigate its emissions (carboNZero 2012).

CarboNZero is a testimony that certification can provide an effective tool in promoting and ensuring carbon mitigation within the voluntary sector. This was acknowledged by the Director of Communications from the United Nations Environment Program, who stated:

CarboNZero is an important example of how a nationally driven greenhouse gas emissions management and certification scheme can advance the global transition to low-carbon economies and societies by encouraging companies, big and small organisations and even individuals to do something about their carbon footprint. Satinder Bindra (carboNZero 2012).

CarboNZero is very transparent on its website[5] in outlining its methodologies and certification process. It clearly asserts that certification is based on international GHG standards and protocols, and it provides a comprehensive list of eligible offsets under its program as well as offset registries where offsets can be retired. This transparency undoubtedly adds credibility to the organisation.

10.4.2 Certified Carbon Neutral Global Standard—The CarbonNeutral® Company

Founded in the UK in 1997, originally under the name 'Future Forests', Carbon-Neutral® is now one of the largest, private international organisations offering Carbon Neutral certification for companies and businesses worldwide. CarbonNeutral® has worked with around 400 companies in over 40 different countries to deliver carbon-neutral certification as well as many other carbon reduction solutions.

CarbonNeutral® certification is based on The Carbon Neutral Protocol (The Carbon Neutral Company 2012), a document, launched in 2002, which provides guidance on how to reach carbon neutrality. The protocol uses international best practice for GHG accounting methods (i.e. WBCSD/WRI's GHG Protocol and Britain's PAS 2060 guidance on carbon neutrality). The certification requires third-party verification to substantiate carbon inventories, footprints and offsetting. It also provides guidance on how it complies with Australia's NCOS.

10.4.3 PAS 2060:2010—Specification for the Demonstration of Carbon Neutrality

A PAS is a *Publicly Available Specification* endorsed by the British Standards Institution (BSI). PAS 2060 was created in 2010 to provide guidance on 'the demonstration of carbon neutrality'. A PAS differs from a formal British Standard in that it is developed relatively quickly by key stakeholders after identifying a lack of information or guidance in an area that is rapidly developing, such as in the case of carbon management and carbon neutrality. The guide is fast-tracked to create a specification under the BSI, and after two years in existence, along with sufficient feedback and revision, a PAS document can become a formal standard under the BSI. [6] As of 2015, this document remains a PAS.

While PAS 2060 provides comprehensive and strict guidance to organisations wishing to achieve carbon neutrality, PAS 2060 and the BSI do not officially certify

---

[5] For more information, see http://www.carbonzero.co.nz.

[6] See http://shop.bsigroup.com/Navigate-by/PAS/.

for carbon neutrality. Compliance is based on self-certification, meaning it is up to organisations to provide enough information to meet the standard. BSI undertakes no official review process. As a result, PAS 2060 demands a high amount of rigour to back up claims. Having developed specific rules around the monitoring of reductions, the purchase and retirement of carbon credits/offsets and the communication of the achievement ensure that it complies with BS EN ISO/IEC 17050-1 standard (British Standards Institution (BSI) 2004).

10.4.4 Carbon Neutral

Carbon Neutral[7] is an Australian not-for-profit organisation providing carbon consultancy, offsets and certification for carbon neutrality. The company was formed by Men of the Trees in 2001, but became its own entity in 2007. Carbon Neutral operates nationally and has become a market leader in providing carbon offsets through biodiverse tree plantings and revegetation.

Carbon Neutral does not officially offer certification under its own logo but can help companies achieve carbon neutrality without any official certification or branding, ensuring that companies do not breach Australia's ACCC Trade Act. Carbon Neutral also provides assistance to companies, products and services wanting to achieve carbon-neutral certification under the federal governments NCOS Carbon Neutral Program.

Currently, any company that donates over 20 tonnes of CO_2 abatement (i.e. 120 trees) can use its logo.

10.4.5 Climate Friendly

Climate Friendly (CF) is an Australian Offset company dealing primarily with the international offset market. As CF's main business is providing offsets, its focus is on ensuring that a high level of rigour and integrity is maintained around this. The carbon footprinting and certification process for businesses wanting to become carbon neutral (or Climate Friendly as its certification states) appear to be secondary to this. Consequently, the carbon footprinting process appears to be less rigorous than other schemes mentioned in this section. Clients wanting to use CF's logo simply enter their emission details into an online carbon calculator, which informs them how many offsets they need to purchase. This lacks transparency in terms of the methodology underpinning the calculations and the process does not require third-party verification.

---

[7] The Australian Carbon Neutral company has no affiliation with the UK's CarbonNeutral® Company.

Similar to carboNZero, Climate Friendly offers two types of certification: one that essentially makes clients carbon neutral ('*Climate Friendly Business*') and another for clients that may only wish to offset part of their total emissions ('*Taking Action With Climate Friendly*') (Climate Friendly 2013).

10.4.6 $NoCO_2$—The Carbon Reduction Institute

The Carbon Reduction Institute was formed in 2006 and offers a variety of carbon management services including carbon-neutral certification within Australia. The company offers several types of carbon certification including $NoCO_2$, $LowCO_2$ and Carbon Neutral. Its methodology is based on recognised international frameworks and protocols but has been adapted, suggesting it is not necessarily comparable or as rigorous as other certification schemes. The certification for carbon neutrality does not specify inclusions/exclusions, but the website states that the 'greenhouse gas impacts produced by and embodied within all products and services consumed' must be accounted for (Climate Reduction Institute 2012).

There is no mention of third-party verification, which is a key component of the international GHG protocols, frameworks and standards, as it provides robustness, rigour and credibility to audits and carbon claims. Without this third-party attestation, this type of certification may prove harder to sell to consumers in the future as people become more critical and demand greater transparency and substantiation to defend the credibility of carbon claims.

10.4.7 Carbon Neutral—ICLEI

As mentioned in Chap. 8, ICLEI has, over several years, been involved in creating various frameworks and protocols to advise cities, local governments and councils on how to measure and manage their carbon emissions and how to achieve carbon neutrality. While councils can claim that they have achieved carbon neutrality under the ICLEI framework, there is no certification process, annual fee or evaluation of documents by ICLEI or another body. However, the guides do encourage participants to seek third-party verification.

10.4.8 NCOS Carbon Neutral Program—Low Carbon Australia

In 2010, the Australian Federal Government introduced an official, government-endorsed Carbon Neutral certification scheme for use in Australia. It appears to be the second[8] of only two government-initiated schemes worldwide. It is based on the NCOS, which has brought credibility and integrity to the Australian voluntary offset market. The NCOS Carbon Neutral Program is the most rigorous, independent and locally relevant scheme for certifying carbon-neutral products and businesses in Australia. It is for this reason that greater focus has been given to understanding this scheme in more depth and to analysing it in terms of it applicability to land development. Further information about the Government NCOS Carbon Neutral Program is provided in Chap. 11.

10.5 Analysis of Existing Schemes

Despite the somewhat limited information available regarding the current state of carbon-neutral certification schemes being offered worldwide,[9] the various carbon-neutral protocols, frameworks and certification schemes mentioned above suggests this is a growing field. The brief analysis above highlights some of the more well-known schemes that currently offer certification for carbon neutrality. The schemes vary in their GHG methodologies, amount of rigour (i.e. requirements for third-party verification), eligible offsets and period of certification. Thus, when it comes to choosing reliable, robust, ongoing carbon-neutral labelling or certification from a reputable (i.e. government endorsed) or internationally recognised organisation, options appear to be limited and issues become apparent.

10.5.1 Vested Interests

One of the primary concerns regarding carbon-neutral certification is the issue of vested interests, particularly in relation to carbon offsetting. A potential conflict of interest can occur when companies who offer carbon footprinting or carbon-neutral certification also supply the offsets. Prospective clients have a right to be dubious of, or question, the carbon footprinting method applied to their organisation, once they understand that it is the offsetting/certification company's intention to sell them the offsets. It would be in the interest of the offsetting company to make the boundaries of the footprint/inventory large if that meant they could sell additional offsets. Similarly, such an organisation has less incentive to encourage rigorous in-house reductions by companies if it decreases their own potential to sell more offsets.

Interestingly, the majority of the above-mentioned certification schemes, despite clear variations in their transparency and rigour, sell offsets as an additional (if not

[8] The first being New Zealand's CarbonZero discussed previously.

[9] No academic studies were found on this specific topic. Research was limited to information available in English.

primary) service. Thus, this can be seen as compromising the integrity or questioning the ultimate motivation of the company providing certification.

Only three organisations do not sell offsets: ICLEI, the BSI (PAS 2060) and Low Carbon Australia. However, the BSI and ICLEI provide only frameworks and standards for achieving carbon neutrality as opposed to actual certification, leaving Low Carbon Australia as the only independent organisation offering purely certification.

10.5.2 Government Endorsement

The 2011 State of the Voluntary Carbon Market Report (Peters-Stanley et al. 2011) lists Australia's NCOS Carbon Neutral Program as one of only five certification schemes and the only government-endorsed program offering carbon-neutral certification. However, the previous section reveals the New Zealand-based 'carboNZero' program, established in 2001, was the first official government led carbon-neutral certification scheme to be implemented globally.

The UK PAS 2060 Standard is also Government endorsed, and this provides guidance on reaching carbon neutrality but does not offer certification. Compliance with PAS 2060 is based on self-certification and organisations making carbon neutral claims under this standard must provide sufficient documentation to support their claim.

The Australian and New Zealand carbon-neutral programs appear to be the only government-initiated programs worldwide offering rigorous, third-party verified carbon-neutral certification.[10] However, as discussed above, the carboNZero program also sells offsets as part of its core business.

10.5.3 A Global Leader

The Australian Government-endorsed NCOS Carbon Neutral Program, with its rigorous GHG accounting methodology, backed up by the third-party verification and linked to a national standard for generating offsets (though separated from the actual selling of offsets), is a very robust, reliable and trustworthy scheme and puts Australia in a unique global leadership position.

It is expected that this Federal Government Carbon Neutral Program will become the accepted standard for carbon neutrality within Australia into the future and perhaps will be seen as global best practice. For this reason, it has been chosen as the primary carbon-neutral certification system to analyse and assess in this book and will be discussed in greater depth in the next chapter.

[10] Though this area is rapidly evolving with new players entering this space frequently.

10.6 Conclusion

This chapter highlighted various issues associated with carbon claims, which arose from the rapid development of the voluntary carbon market and associated carbon industries. These issues led to mounting credibility and validity issues around the quality of services being offered and the claims being made. In Australia, this resulted in the introduction of new laws and fines to discourage companies from making false and misleading carbon claims.

Certification schemes for carbon neutrality emerged as a way to increase the integrity of carbon claims and thus increase consumer confidence in them. Some of the well-known Australian and international carbon certification schemes were identified and discussed.

While none of the schemes have a specific process for certifying urban development, the Australian NCOS Carbon Neutral Program appears to be one of the most rigorous existing certification schemes and is thus discussed in greater detail in the next chapter, including an analysis of the scheme's ability to certify land development.

References

Australia Competition and Consumer Commission (ACCC). (2008). *Carbon claims and the trade practices act*. Retrieved from http://www.greenwashreport.org/downloads/ACCC_Carbon.pdf.

Brecard, D., Hlaimi, B., Lucas, S., Perraudeau, Y., & Salladarre, F. (2009). Determinants of demand for green products: An application to eco-label demand for fish in Europe. *Ecological Economics, 69*, 115–125.

British Standards Institution (BSI). (2004). BS EN ISO/IEC 17050-1:2010 *Conformity assessment—Supplier's declaration of conformity—Part 1: General requirements*. Available at http://shop.bsigroup.com/ProductDetail/?pid=000000000030095948.

carboNZero. (2012). *Testimonials*. Retrieved from http://www.carbonzero.co.nz/why/testimonials.asp.

Climate Friendly. (2013). *Climate friendly logo usage*. Retrieved from http://www.climatefriendly.com/Business/ResourceCentre/LogoUsage/.

Climate Reduction Institute. (2012). *$NOCO_2$ certification program: Scope and methodology*. Retrieved from http://www.noco2.com.au/web/page/scope_methodology.

Hamilton, S. F., & Zilberman, D. (2006). Green Markets, eco-certification and equilibrium fraud. *Journal of Environmental Economics and Management, 52*, 627–644.

International Organisation for Standardisation (ISO). (2004). ISO/IEC 17000:2004 *Conformity assessment—Vocabulary and general principles*. Retrieved from http://www.iso.org/iso/catalogue_detail.htm?csnumber=29316.

International Organisation for Standardisation (ISO). (2013). Standards. *Certification*. Retrieved April 24, 2013, from http://www.iso.org/iso/home/standards/certification.htm.

Jamalpuria, A. (2012). On voluntary eco-labeling and fiscal incentives. *Economic Letters, 117*, 110–114.

Murray, J., & Dey, C. (2008). The carbon neutral free for all. *International Journal of Greenhouse Gas Control, 3*, 237–248.

Organisation for Economic Co-operation and Development (OECD). (2002). *Frascati manual: Proposed standard practice for surveys on research and experimental development: The Measurement of Scientific and Technological Activities*. Organisation for Economic Co-Operation and Development.

Peters-Stanley, M., Hamilton, K. Marcello, T. & Sjardin, M. (2011). *Back to the future: State of the voluntary carbon markets 2011: Ecosystem marketplace*. Bloomberg New Energy Finance.

The Carbon Neutral Company. (2012). *The carbon neutral protocol: The global standard for carbon neutral programmes*. Retrieved from http://www.carbonneutral.com/interface/files/aboutus/qualityassurance/TheCarbonNeutralProtocol.pdf.

Chapter 11
Spotlight: The Australian Government Carbon Neutral Standard

11.1 Introduction

The move towards national carbon pricing in Australia in 2009 led to a re-evaluation of the existing voluntary carbon offset market and what it means to become carbon neutral within an Australian context. These policy developments led to the establishment of a new National Carbon Offset Standard (NCOS), which ensured that offsets bought and sold—thus, inherently connected to carbon neutrality—remain additional to those carbon reductions expected to occur under a regulated carbon price.[1]

Alongside this overarching national standard for carbon offsetting, the government introduced a rigorous, government-endorsed process for certifying carbon neutral products, services and organisations—the NCOS Carbon Neutral Program (CNP). This programme has helped to provide clarity and credibility to a rapidly growing field and industry in Australia.

Shortly after the release of the NCOS CNP in 2010, the Australian Federal Government created an independent body called Low Carbon Australia Limited (LCAL) to be responsible for administering the certification process. The certification process carried out by LCAL was an independent, stand-alone service, i.e. GHG audits, third-party verification and the purchase of carbon offsets each remain a separate exercise, providing LCAL with greater impartiality and credibility. In 2014, after a change in government, LCAL merged with the Clean Energy Finance Corporation (CEFC), and the administration and management of the NCOS CNP shifted to the newly formed Department of Environment.

The section below begins with an overview of the NCOS, as it plays a pivotal role in the certification process, before examining the specific process for achieving

[1] Australia had a mandatory carbon price on large emitters from 2012 to 2014. It was repealed after a change in government in 2013.

V. Rauland and P. Newman, *Decarbonising Cities*, Green Energy and Technology, DOI 10.1007/978-3-319-15506-7_11

carbon neutrality under the NCOS CNP. Some issues arising around the certification of land development under this programme are identified and discussed, as well as the impact that the change in government and the repeal of the carbon pricing legislation has had on the NCOS CNP.

11.2 The National Carbon Offset Standard

In 2009, the Australian Federal Government developed the NCOS to increase the credibility of, and consumer confidence in, carbon offset claims in Australia. It was also designed to bring both consistency and reliability to the Australian voluntary carbon offset market in the light of the proposed new Carbon Price Mechanism (CPM). By ensuring that offsets were additional to Australia's national targets and then planned CPM,[2] businesses and organisations could be assured that their voluntary actions and commitments demonstrated a genuine contribution to Australia's national emissions reduction effort (Climate Group 2009).

11.2.1 Eligible Offsets Under the NCOS

There are various types of offsets currently accepted under the NCOS, including both domestic and international offsets. International offsets consist of either Kyoto units or voluntary units from a range of different schemes discussed in Chap. 7 and outlined in Table 11.1.

The types of eligible offsets have changed since the repeal of the country's carbon price in 2014. There is now only one source of domestic offset units eligible under the NCOS: Australian Carbon Credit Units (ACCUs), which are developed through the Federal Government's Carbon Farming Initiative (CFI), which is transitioning to the new Emission Reduction Fund (ERF)[3] (DCCEE 2012a, b).

The CFI is the only scheme[4] currently generating offset credits from projects within Australia. Prior to the repeal of the carbon price, these domestic offsets had to come from activities from sectors outside those covered by the CPM (e.g. agriculture). However, the repeal of the mandatory price has opened the door for a variety of previously ineligible sectors to now participate in the development of voluntary credits/offsets. Some of the new sectors include the following:

---

[2] The Carbon Price Mechanism came into force in July 2012 and was later repealed in June 2014.

[3] Carbon units, which were generated through the mandatory Carbon Pricing Mechanism, would have been eligible as voluntary credits after 1 July 2015, though as the CPM legislation has been removed, these credits are no longer generated. Greenhouse-friendly credits were also available until July 2013.

[4] Credits generated under the CFI can be purchased by the government through the ERF scheme.

Table 11.1 Eligible offsets under the NCOS

	Type of offset	Name of unit	Program
Domestic offsets	Australian government units	Australian Carbon Credit Units (ACCUs)	Carbon Farming Initiative (CFI)
International offsets	Kyoto units	Certified emissions reductions (CERs)	Clean Development Mechanism (CDM)
		Emissions reduction units (ERUs)	Joint Implementation (JI)
		Removal units (RMUs)	Land use, land-use change and forestry (LULUCF) activities
	Voluntary units	Verified carbon units (VCUs)	Verified carbon standard
		Voluntary emissions reductions (VERs)	Gold standard

- Efficiency improvements in the commercial building sector;
- Waste coal mine gas;
- Cleaning up power stations;
- The transport sector; and
- Large industrial facilities. (Clean Energy Regulator 2014)

Projects applying for carbon credits through the CFI must use an approved carbon abatement methodology. Methodologies can be developed by private applicants or industry associations. Methodologies are approved by the Emissions Reduction Assurance Committee (ERAC), previously known as the Domestic Offsets Integrity Committee (DOIC). These committee's were set up as independent committees to provide expert advice and to assess the abatement methodologies of offset projects submitted to the government. Currently, 26 methodologies have been approved and are broadly grouped under agriculture (e.g. livestock, soil carbon, fertilisers, feral animals), vegetation projects (e.g. environmental plantings, savannah burning, afforestation and reforestation), and landfill and alternative waste treatment projects (e.g. landfill methane capture and combustion, and landfill diversion).[5] With so many new sectors now allowed, it is expected that many more methodologies will emerge in the coming years. Table 11.1 summaries the offsets eligible under the NCOS.

To become certified as carbon neutral under the NCOS CNP, only offsets recognised under this standard are able be used. It is important to note that offsets can still be generated from a range of other activities and schemes that are not recognised by the NCOS. While these offsets cannot be used under the Federal

[5] As of the 27 December 2014, See http://www.environment.gov.au/climate-change/emissions-reduction-fund/cfi/methodologies/determinations.

Government's NCOS CNP, they can be used to make carbon neutral claims outside of the programme, so long as the claim (including information about the type of offset) is clearly explained to satisfy the requirements of the ACCC.

11.3 NCOS Carbon Neutral Program and Certification Process

The NCOS CNP commenced in 2010, marked by the release of the first version of the NCOS Carbon Neutral Program Guidelines (CNPG). As mentioned, the NCOS CNP is underpinned by the NCOS and based largely on the GHG Protocol developed by the World Resources Institute and World Business Council for Sustainable Development (WRI and WBCSD 2004) and Australia's National Greenhouse and Energy Reporting Systems (NGERS) (Clean Energy Regulator 2014) and is consistent with the ISO documents (ISO 14064 series) outlined in Appendix C. Certification follows the broad three-step process outlined in Chap. 7, which involves first calculating the carbon footprint of the product or organisation, then reducing emissions where possible or financially viable, and finally purchasing eligible carbon offset credits to offset the remaining unavoidable emissions.

The NCOS carbon neutral certification runs over a five-year period and requires:

- annual reporting of emissions and updating of the emissions management plan;
- biannual third-party verification of the audit;
- annual purchasing of offsets; and
- payment of an annual certification fee.

The general process for achieving carbon neutrality is highlighted in Fig. 11.1.

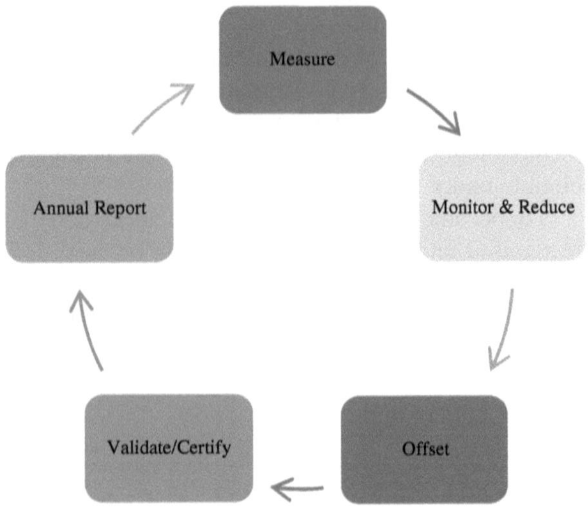

Fig. 11.1 Overview of the carbon neutral process required by LCAL

The licence fee structure is determined by the size of the footprint being assessed.

There are four key documents that must be submitted annually as part of the certification process. These are as follows:

1. The Greenhouse Gas Inventory/Carbon Footprint;
2. The Emissions Management plan (EMP);
3. The Public Disclosure Summary (PDS); and
4. The Annual Report.

These documents are briefly described in the section below.

11.3.1 The Greenhouse Gas Inventory/Carbon Footprint

The NCOS and CNPG allow participants to choose between the two types of carbon footprint calculation methodologies mentioned in Chap. 7 (and further outlined in Appendix A). These include either the inventory approach (targeting organisations) or a life cycle analysis approach (recommended for products and services). However, after a period of public consultation, the revised CNPG also permits and suggests using a hybrid approach for calculating the carbon footprint of certain projects such as construction works, which may not fall specifically within either category.

The applicant then determines the boundaries of the organisation, deciding, for example, whether the entire operation is to become carbon neutral or only a section (i.e. a department or aspect of the business).[6] Emission sources and Scopes are affected by the methodology and boundaries chosen. Nevertheless, the guidelines require that all Scope 1 and 2 emission sources be included and recommend where feasible, that all Scope 3 sources also be incorporated. This is far greater than what is required under the NGERS reporting, particularly for small organisations, where Scope 3 emissions often represent a far greater proportion[7] of total emissions (Downie and Stubbs 2011).

11.3.1.1 Scope 3 Emissions

Recognising the magnitude of, and complexity associated with Scope 3 emission sources, the GHG Protocol released a new standard in 2011 called *Corporate Value Chain (Scope 3) Accounting and Reporting Standard* (WRI and WBCSD 2011). Table 11.2 provides a summary of the framework for identifying the various Scope

[6] It is generally too complicated to separate a product's boundaries, particularly when marketing, hence the requirement for an LCA methodology.

[7] Downie and Stubbs (2011) suggest it can be up to 75 % of a company's total emissions.

Table 11.2 Scope 3 emissions. Adapted from WRI and WBCSD (2011)

Potential Scope 3 emissions	
'Upstream emissions'	Purchased goods and services
	Capital goods
	Fuel- and energy-related activities (not included in Scope 1 or Scope 2)
	Upstream transportation and distribution
	Waste generated in operations
	Business travel
	Employee commuting
	Upstream leased assets
'Downstream emissions'	Downstream transportation and distribution
	Processing of sold products
	Use of sold products
	End-of-life treatment of sold products
	Downstream leased assets
	Franchises
	Investments

3 emissions proposed by this new protocol. The sources are identified as either 'upstream' Scope 3 emissions or 'downstream' Scope 3 emissions (WRI and WBCSD 2011).

While the CNPG recommend including as many Scope 3 emission sources as possible, they require at a very minimum that the Scope 3 emissions for paper, waste and business travel be accounted for in the carbon footprint (DIICCSRTE 2013). The latest version of the NCOS CNP guidelines, however, acknowledge that these stipulations are not necessarily relevant for all carbon neutral applications (i.e. construction and land development)[8] and thus advocate an individual appraisal and identification of the most relevant and feasible sources of Scope 3 emissions.

11.3.1.2 Renewable Energy

The purchase or use of renewable energy (or GreenPower™) in relation to carbon claims can be interpreted in many ways. In order to avoid confusion, the NCOS CNP developed specific rules relating to the use of GreenPower™. As such, where government-approved GreenPower™ has been purchased from an electricity retailer, the emissions associated with this electricity consumption (i.e. an organisation or product's Scope 2 emissions) will equal zero. A similar situation exists if a company produces its own renewable electricity onsite, that is the emissions from

[8] This is the only mention of the construction sector in this revised proposal. As yet, no construction company or developer has sought certification.

this electricity will simply equal zero. However, it must provide evidence to the Department of Environment that the Renewable Energy Certificates (RECs) associated with the renewable electricity generation have been retained by the owner and cancelled and not sold on.[9] Generated or purchased electricity from renewable energy sources cannot be used to offset other internal emissions associated with an organisation or product's carbon footprint under the NCOS CNP. This poses a problem for land developers, who want to produce more energy than they consume. This will be discussed later.

11.3.2 The Emissions Management Plan

The Emission Management Plan (EMP) underpins and documents the entire carbon neutral process under the NCOS CNP. The EMP is defined within the CNPG as an outline of 'the management framework, systems and processes … in place to effectively manage your carbon neutral commitment from start to end' (DCCEE 2011b). The EMP must provide information on the greenhouse gas audit, data collection and record keeping, quality control practices, the emissions reduction strategy (which highlights all emission abatement activities and initiatives including justification of all reduction estimates), the carbon offset purchasing process, and verification and identification of uncertainty within the audit.

11.3.3 The Public Disclosure Summary

The Public Disclosure Summary (PDS) extracts the main points from the EMP (often a lengthy document) to create a short, one to two page summary highlighting the key areas of the applicant's process to demonstrate to the public how the applicant achieved carbon neutrality. The purpose of this document is to promote transparency and credibility of the certification process, thereby strengthening the carbon claim.

11.3.4 Annual Report

As mentioned, the documents must be updated and submitted annually to provide evidence to the Department of Environment that the applicant is on track in meeting its NCOS obligations under the CNP. Every second year, the audit and process must be reviewed by independent, third-party, NGERS qualified GHG verifiers.

---

[9] If RECs are sold, the electricity produced is no longer considered carbon-free as the certificates are going towards assisting a liable entity to meet their quota under the Federal Government's 20 % Renewable Energy Target (RET). If RECs are retained by the producer of the electricity (e.g. from rooftop PVs), this ensures Australia goes above and beyond what the RET mandates.

The annual report must be submitted within four months of the end of the reporting period. This has significant implications for urban development and is thus discussed below.

11.4 Issues for Certifying Precinct-Level Urban Development Under NCOS

Vast amounts of literature over several decades have highlighted the impact the construction industry has on the environment and greenhouse gas emissions (ASBEC 2008; Cole and Rousseau1992; Dixit et al. 2012; Zuo et al. 2012). Developers are increasingly becoming aware of their environmental impact, and many are taking proactive measures to reduce it. Several are striving to become leaders in their field and differentiating themselves in the market based on their green credentials.[10]

Carbon neutral certification provides one way of demonstrating this environmental commitment to the public. Striving for carbon neutrality also provides numerous other benefits, as mentioned in Chap. 7, such as the reduced long-term operating costs of living and working in low-carbon precinct developments (e.g. reduced energy, water, waste and transport costs), increased health benefits and increased productivity, amenity and liveability.

The Australian Federal Government's NCOS Carbon Neutral Certification Program has provided a world-class platform for acknowledging the carbon reduction achievements of organisations in Australia. Being underpinned by robust, international standards for carbon accounting and overseen by leaders in the field, this programme is pioneering its way through a complex and unique set of conditions in relation to Australia's carbon market, which is set within a capricious political landscape. With these complex circumstances in mind, and the fact that the NCOS CNP is still relatively in its infancy, it is not surprising that various issues are still being identified and worked through.

When applying the NCOS CNP certification process to precinct-scale urban development, several issues become evident. To begin with, the generic guidelines mean that many decisions, such as those about boundaries or ways of dealing with various types of developments (e.g. new development versus redevelopment), are relatively open to interpretation, making comparisons less robust and effective. There is currently little guidance in the carbon neutral certification process around how to manage the multiple stakeholders that are inevitably part of larger, precinct-scale developments. Ongoing certification poses another problem for developers who often hand over responsibility once a development or parts of it are sold.

Offsets are yet another challenge for developers wishing to attain carbon neutral certification through the NCOS CNP, as many of the onsite emissions reduction

---

[10] See the pixel building by Grocon http://www.grocon.com.

opportunities within urban development are not considered eligible offsets under this scheme. How to deal with behaviour change in the long term and how to ensure that the calculation process and transaction costs associated with certification do not become too onerous or prohibitive are additional areas that need to be considered. These areas are discussed in more depth below.

11.4.1 Generic Guidelines

The CNPG are intentionally generic in order to be relevant to the largest possible number of participants. However, having generic guidelines may lead to inconsistent GHG methodologies being adopted within certain subsectors such as precinct-scale development. For example, the current guidelines, which primarily target organisations and/or their products, recommend either the inventory approach or a life cycle analysis for calculating emissions. However, as the construction of urban precinct development is more consistent with an ongoing project than an organisation or a typical product, it tends to fall between the two proposed methodologies. The revised guidelines (DCCEE 2011b) recommend that in circumstances where there is no clear methodology, such as for construction, a hybrid approach can be adopted. However, this could be seen as further exacerbating the problem of inconsistency and incomparability in terms of reporting.

The hybrid approach may well be the best option for calculating the footprint of a precinct-scale development, but a consistent definition relating specifically to the boundaries of the development needs to be in place. Identifying and defining these boundaries remains the biggest challenge in achieving consistency of reporting. Kennedy and Sgouridis (2011) in their article 'Rigorous classification and carbon accounting principles for low and Zero Carbon Cities' acknowledge the difficulty in defining these boundaries, pointing out that the city with its precincts is a:

> dynamic and complex system, defined in part by its geographic boundaries, but also by its interconnections with a much broader region through exchanges of materials, energy and information. This interconnectedness complicates the task of determining which emissions should be included. (5261)

With this in mind, it is argued that sector-specific guidelines are needed to bring greater consistency to the reporting of emissions of developments at the precinct level to ensure that, at a minimum, certain key emission sources such as the six highlighted in this book[11] are included in the emissions calculation, regardless of the chosen methodology.

[11] i.e. the embodied emissions in the materials of buildings, roads and infrastructure, and the upstream and downstream emissions associated with the energy, water and waste cycles and transport.

11.4.2 Catering for Different Types of Development

Another factor that is likely to have a significant effect on the overall footprint of a development is the type and location of the development. There are three broad classifications of urban development. These are 'greenfield' developments (sites where no previous development has taken place), 'greyfield' developments (previously light industrial/commercial or older residential sites) and 'brownfield' developments (previously heavy industrial sites, often requiring land remediation). The titles broadly establish whether a development is using reclaimed land (a greyfield or brownfield redevelopment site) or is a brand new development, often situated at fringe of the city (a greenfield site)—see Newton et al. (2012).

The potential variation in footprints is likely to be significant when factors such as occupant transport and the embodied emissions associated with construction are considered. For example, if a project reuses materials already onsite from a previous development, or if aspects of an existing development are retained and reconstructed, the embodied emissions are considerably less (e.g. if there is retention of foundations, the building's core, or water and gas pipes). In this case, it should be stipulated in the carbon accounting methodology or guidelines that only the embodied carbon that is new to the site be included in the carbon footprint.

Kennedy and Sgouridis (2011) support this idea, suggesting that the embodied emissions included in a carbon footprint of a redevelopment be based on the starting year of a carbon claim, thereby allowing emissions already locked up in existing buildings and infrastructure be excluded from the audit. Care will be needed to ensure that this doesn't disincentivise developers from taking action at the design stage.

The carbon footprint of a development is ultimately defined by its location and connection to low-carbon transport infrastructure. Although greyfield or brownfield sites can still vary greatly in respect with their accessibility to public transport, these sites typically have fewer transport emissions than greenfield sites,[12] as they are located in more established areas, often with amenities in a closer proximity (Newman and Kenworthy 2015).

These considerations should be taken into account when designing guidelines to properly acknowledge and reward genuine low-carbon and carbon neutral developments. It stands to reason that a development, no matter how energy efficient it is or how many sustainability features or technologies it has incorporated, cannot be truly carbon neutral if the residents and businesses are forced to use emissions intensive car-based transport to commute long distances to and from the development.

[12] With the exception being if a new greenfield development is being built as a transit-oriented development.

11.4.3 Dealing with Multiple Stakeholders

Another pertinent issue for developers aspiring to gain carbon neutral certification involves managing multiple stakeholders. There are countless stakeholders associated with precinct-scale development, including developers, architects, urban planners, local councils, state government agencies, energy, water and waste utilities, construction workers and tradesmen, and existing and potential residents and businesses. Each stakeholder has an ability to affect, either positively or detrimentally, the outcome of the development and attainment of carbon neutral status. Blundell (2012) specifically highlights the importance of the utility sector as key stakeholders within precinct-scale development, suggesting that without their commitment, collaboration and support, sustainability outcomes will fail.

In order to be successful in achieving ambitious sustainability goals, a developer needs the commitment and ongoing support from all stakeholders (e.g. utilities, urban planners, architects and tradespeople). Stakeholders therefore need to be engaged in the process from the beginning, and the aspirations for the development (i.e. carbon neutrality) need to be made clear from the outset of the project. Participation and support are likely if there is a perceived benefit to all parties (Thabrew et al. 2009). In a participatory governance exercise in Geraldton, Western Australia, (Hartz-Karp 2012; Gollagher and Hartz-Karp 2013) found that carbon reduction projects and carbon neutrality received much higher importance in their final infrastructure decisions than if the process had not been participatory.

Establishing a clear understanding of the costs and benefits associated with the project on each of the stakeholders is critical. The notion of 'split incentives'[13] is likely to be an issue for urban development striving for carbon neutrality. Split incentives involves a situation where a set of conditions (i.e. costs and risks) prevents certain parties (e.g. owners) from investing in improvements or upgrades in a building if the benefits accrue to a different set of stakeholders (e.g. tenants) (ASBEC 2008; DCCEE 2010; WBCSD 2009). If left unresolved, this can lead to inertia and stagnation and prevent the attainment of goals.

In terms of carbon neutral development, the question will be: Who ultimately benefits from the carbon neutral development, in particular, from certification, and who pays the cost of achieving and maintaining it? To ensure all stakeholders share in the benefits, as well as the associated costs and risks, these factors will need to be identified and worked through at the beginning of the process.

There are some examples where good stakeholder participation has led to exceptional results. The successful design and development of Vauban eco-community is one such project. Having encouraged strong community and stakeholder participation from the outset, the ideas, aspirations and goals generated throughout the process were essentially owned by all (Williams 2012). Thabew et al. (2009) also acknowledges the importance of community, suggesting that they should be at

[13] Also known as 'Principal-Agent problem' (DCCEE 2010) and 'Landlord-Tenant problem' (ASBEC 2008).

the centre of the process to ensure that they, along with other key stakeholders, feel 'empowered to influence and share control over development initiatives, decisions, and the resources affecting them' (67–68).

Examples such as Vauban suggest that an integrative, participatory sustainability planning approach involving multiple stakeholders is likely to increase the likelihood of success in achieving ambitious or complex goals within urban development, yet despite this, formal methodologies for managing the process still appear limited (Thabrew et al. 2009). Yang et al. (2011) suggest there is no universally accepted structured approach for engaging and managing stakeholders within the construction industry, though several methodologies and approaches do exist for managing and facilitating stakeholder engagement processes in general (Reid 2008; Thabrew et al. 2009).

11.4.4 Issues with Ongoing Certification

While the need for ongoing certification of carbon neutrality has previously been identified as a key requirement to ensure carbon claims remain valid and accurate, there are likely to be several issues associated with maintaining this in relation to urban development. The primary issue concerns ownership of the development. Developers rarely maintain control or ownership over urban development projects once the construction is completed. Thus, the question that arises is: Who will be responsible for ensuring that ongoing measurement, monitoring, emissions reductions and purchasing and retiring of offsets will be carried out in line with the requirements under the NCOS CNP?

Once land development is sold off to individual landholders (both commercial and residential), it will become increasingly challenging to obtain agreement among all new owners to maintain certification.[14] However, if by chance, the decision to pursue and maintain certification *is* agreed upon by the various stakeholders, the question still remains: Who will be responsible for undertaking the process (i.e. conducting the annual carbon footprinting and writing up emissions management plans) and who will be responsible for implementing measures and ensuring that targets are being met/enforced?

The outcome is likely to be that land developments may only seek certification in their design phase. Finding further emissions reduction opportunities within an urban development that has been designed well will also prove challenging as the majority of emissions reduction measures will have been implemented in the design phase. Nevertheless, new technologies are continuously emerging and becoming available, providing new emissions reduction opportunities into the future, so the possibilities

[14] This issue also impacts on the commercial building rating schemes that have proliferated around the world. While these tools are primarily design tools, attempts are being made to create operational ratings. Nevertheless, the mix of owners and users in large buildings usually makes it difficult to seek any ongoing certification or rating.

of an operational certification scheme should not be abandoned. New governance structures will be required to enable this. These are discussed further in Sect. 12.3.3.

11.4.5 Offsets

Currently, there is no specified limit on the number of credits or offsets that can be purchased or used within a development when claiming carbon neutrality under the NCOS CNP. The only requirement is that applicants demonstrate that onsite carbon reductions are also being pursued. The main rules, requirements and guidelines around offsets focus on the *types* of offsets allowable under the standard.

As mentioned in Chap. 7 and Appendix A, surplus carbon-free renewable energy generated onsite within a development and fed into the grid is commonly seen as a form of offsetting. When applied to buildings, this concept is referred to as 'net-zero energy', whereby the buildings feed as much energy generated onsite into the grid as the building has taken from the grid over a year period (Hernandez and Kenny 2009).[15] While this process has traditionally not taken into account the embodied emissions associated with buildings or infrastructure, Hernandez and Kenny (2009) argue that it should and propose a new definition of zero energy from a life cycle perspective as follows:

> ...the primary energy used in the building in operation plus the energy embodied within its constituent materials and systems, including energy generating ones, over the lifetime of the building is equal to or less than the energy produced by its renewable energy systems within the building over its lifetime. (817)

Williams (2012) suggests that developments that claim to be net-zero carbon on an annual basis can do so by using this process. However, this concept and method of balancing or neutralising emissions is not accepted under the Australian Federal Government's NCOS CNP, which is based on internationally recognised protocols (DCCEE 2011b). One of the key eligibility requirements for offsets under these protocols is that they meet the additionality criteria mentioned previously—that is, that offsets are additional forms of abatement relative to those which would occur under a BAU scenario (WRI and WBCSD 2004).

The reason why energy-related projects were not seen as eligible offsets was due to issues with the additionality criteria under the mandatory CPM. As energy generation was a covered sector under the previous Australian Government's national carbon pricing legislation (which ran from 2012 to 2014), developers were already paying the equivalent of a carbon tax on their energy-related emissions, either directly or as a cost passed on from energy utilities. Therefore, as developers were already fiscally encouraged to reduce carbon, it could be argued that any action to further cut emissions from energy production or use onsite would have occurred as a result of the legislation. Consequently, energy-related offsets were not deemed 'additional'

[15] Kennedy and Sgouridis (2011) refer to this as 'balancing'.

voluntary forms of abatement under the NCOS. The only eligible domestic offsets under the legislation came from the CFI and remaining greenhouse-friendly offsets[16] (DCCEE 2011a, b, 2012a, b). This has changed now with the repeal of the carbon price legislation and the introduction of the Emission Reduction Fund (ERF) and Direct Action Plan (DAP). International offsets were/are also eligible.

Therefore, up until 2014, the only fully articulated guidelines and certification scheme in Australia (i.e. the NCOS CNP), which could potentially recognise a precinct-scale development's carbon reduction, contained a rule that prevented onsite carbon reduction from energy being recognised as an offset. This issue would make it particularly difficult (i.e. expensive) for developers to achieve carbon neutrality using the eligible offsetting methods.

So in a time where integrated renewable-based urban development is being pushed by many (e.g. Droege 2006) and is emerging as a major part of the urban landscape in Australia (Newton and Newman 2013), urban developers in Australia were being driven away from using onsite renewables and other innovative energy efficiency projects to reduce emissions. Instead, they were being steered towards using offsets that are not seen to be part of the development, but just an accounting process. Developers and prospective buyers or renters were therefore not being encouraged to create the kind of eco-city developments that are currently becoming global best practice in green urban development.

This was a major flaw in the certification system. However, much of this has changed when the Carbon Price Mechanism was repealed in 2014.

11.4.6 Australia's Direct Action Policy and its Effect on Offsets

The 2013 federal election in Australia resulted in a change in government, from one that was recognised globally as taking significant action on climate change, to a government, which believed the market-based policies in place impacted far too much on household budgets and industry competitiveness. While the government acknowledges human-induced climate change[17] and remains committed to meeting the same emission reduction targets as the previous government, their chosen mechanism for achieving the target is generally seen as inferior.

While in opposition, the coalition parties made an election promise to remove the CPM. Nine months after the coalition government was elected, in July 2014, the CPM was successfully repealed. The new policy that replaced the CPM was known

[16] Greenhouse-friendly offsets were only eligible until July 2013.

[17] Though the Prime Minister, Tony Abbott, is on record stating he personally does not believe in human-induced climate change, or see sufficient reason to take action which will place a financial burden on the economy. See http://www.theguardian.com/environment/planet-oz/2014/jun/16/what-does-australian-prime-minister-tony-abbott-really-think-about-climate-change.

as the DAP and its central policy is the ERF. At the ERF's current level of funding, it is anticipated to fall short of the required carbon abatement required to meet the targets. However, the DAP did introduce greater flexibility in some areas and reduced the time requirement for permanence in the land use and forestry sector. It also effectively removed the aforementioned issue associated with double counting and additionality for the land development sector, opening up new possibilities for developers to be recognised for a variety of carbon reduction initiatives.

The DAP allows many more sectors to participate in the generation of carbon credits, than were eligible in the presence of a carbon price. Project proponents still need to have their methodologies assessed by the ERAC. Projects using approved methodologies will then be assessed by the Clean Energy Regulator (CER) to determine if they can be registered. There are various requirements that need to be met at this stage. Once a project has been approved by the CER, it is eligible to tender its anticipated abatement to government at the expected cost. Projects are chosen on their ability to provide carbon abatement at least cost. They are selected through a reverse auction process, whereby the least cost projects are chosen first. Successful projects are funded through the ERF.

ACCU's are administered on delivery of abatement (i.e. at the end of each year or when reductions are actually recorded), which are then bought back by the government through the ERF.

As mentioned in Sect. 11.2.1, the DAP has enabled energy efficiency and other previously ineligible carbon reduction initiatives and projects to be eligible to receive carbon credits (provided approved methodologies are used). In theory, if developers retire ACCU's generated from their development instead of selling them back to the government, these ACCU's could potentially be used to offset other emissions associated with their development under the NCOS CNP.

11.4.7 Behaviour Change

A big issue with any new green building development or new technology relates to operational behaviour. Assessment of the BedZED development showed that people who came into the development without prior knowledge were able to save carbon about half the national carbon footprint (due to physical design elements), but those educated and committed to reducing their carbon footprint were able to save considerably more (Newman and Jennings 2008). If residents or workers within a low-carbon environment are not educated about how to live or work sustainably, then the goals and targets are unlikely to be fully reached. Nevertheless, this does not discount the fact that design can still achieve worthwhile carbon benefits.

Table 11.3 Low-carbon Australia 2012/13 fee structure

Tire	Level of emissions	Price
1	<2000 tCO_2-e	$3230
2	2000–10,000 tCO_2-e	$9,631
3	>2000 tCO_2-e	$22,773

Source P. Watt (Personal Communication 13, May, 2013)

11.4.8 Transaction Costs

Transaction costs are another factor that needs to be considered when deciding whether to seek carbon neutral certification. There are several different fees associated with gaining certification through the NCOS program. These include fees for the administration of the certification, which are paid to Department of Environment, fees to an independent third-party verifier and the costs associated with purchasing offsets. In addition, developers with no prior experience in conducting a carbon footprint are likely to need to hire a consultant to undertake this initial work (which is then verified by a third-party consultant).

Currently, there are three tiers of certification fees based on level of emissions the applicant is responsible for. The 2013 fees are provided in Table 11.3.

The costs associated with undertaking the initial carbon footprint of a development and third-party verification can vary considerably depending on the organisation carrying it out. While it is difficult to estimate the cost for these services, as this depends on the size of the development, it is expected to be higher than a typical organisational footprint. Furthermore, as large industry, including the mining and energy sector, currently dominates the demand for carbon accounting, costs for these services are relatively high and unrealistic for smaller organisations. This will most likely change as demand for carbon accounting and green jobs grows.

Although offsets are likely to be the smallest cost associated with the carbon neutral certification process[18] at present, it is expected that the costs of offsets will rise in the future in line with a rising global carbon price.

The transaction costs associated with obtaining certification are not insignificant and may deter some applicants from pursuing this as an option.

11.5 Conclusion

Although the NCOS CNP is identified as one of the leading independent certification schemes worldwide, a deeper analysis has revealed several issues when applying the certification scheme to urban development. These included generic guidelines which fail to provide sufficient guidance to developers, how developers

[18] Based on the current availability of cheap international offsets.

deal with a variety of stakeholders at the precinct level, and how they deal with ongoing certification, the types of eligible offsets that discourage onsite renewables, behaviour change issues (as ongoing certification for urban development is predominantly about the people living and working there) and the costs associated with certification. Further investigation is needed to identify opportunities for dealing with the various issues identified, which may require changes to the NCOS CNP guidelines. These are dealt with further in Chaps. 12 and 13. The introduction of the Australian Federal Government's End-User focussed Direct Action Policy may assist in solving some of these issues, particularly around offsets. The potential is there for some demonstrations of this kind of mainstreaming, which could then be applied anywhere.

References

Australian Sustainable Built Environment Council (ASBEC). (2008). *The second plank—building a low carbon future with energy efficient buildings*. Retrieved from http://www.asbec.asn.au/files/ASBEC%20CCTG%20Second%20Plank%20Report%202.0_0.pdf.

Blundell, L. (2012). *Environmental upgrade agreements*. The Fifth Estate. Retrieved from http://www.thefifthestate.com.au/wp-content/uploads/2012/10/TFE_eBOOK_FINAL-41.pdf.

Climate Group. (2009). *The role of voluntary action in light of the CPRS and Australia's ratification of the Kyoto Protocol*. Retrieved from http://www.theclimategroup.org/_assets/files/The-role-of-voluntary-action-in-light-of-CPRS—May-2009.pdf.

Cole, R., & Rousseau, D. (1992). Environmental auditing for building construction: Energy and air pollution indices for building materials. *Building and Environment, 27*(1), 23–30.

Clean Energy Regulator. (2014). *National greenhouse and energy reporting guidelines*. Retrieved from http://www.cleanenergyregulator.gov.au/National-Greenhouse-and-Energy-Reporting/Pages/default.aspx.

Department of Climate Change and Energy Efficiency (DCCEE). (2010). *Report of the Prime Minister's Task Group on Energy Efficiency*. Retrieved from http://www.industry.gov.au/Energy/EnergyEfficiency/Documents/energy-efficiency/workshops-report.pdf.

Department of Climate Change and Energy Efficiency (DCCEE). (2011a). *Greenhouse friendly*. Retrieved from http://www.climatechange.gov.au/greenhousefriendly/.

Department of Climate Change and Energy Efficiency (DCCEE). (2011b). *Carbon neutral program guidelines: National carbon offset standard*. Version 2.0. Retrieved from http://www.climatechange.gov.au/sites/climatechange/files/documents/03_2013/cnp-guidelines.pdf.

Department of Climate Change and Energy Efficiency [DCCEE]. (2012a). *An overview of the carbon farming initiative*. Retrieved from Department of Climate Change.

Department of Climate Change and Energy Efficiency (DCCEE). (2012b). *National carbon offset standard. Version 2*. Retrieved from http://www.climatechange.gov.au/ncos.

Department of Industry, Innovation, Climate Change, Science, Research and Tertiary Education (DIICCSRTE). (2013). *Australian national greenhouse accounts: National inventory report 2011*. Retrieved from http://www.climatechange.gov.au/sites/climatechange/files/documents/05_2013/AUS_NIR_2011_Vol1.pdf.

Dixit, M. K., Fernández-Solís, J. L., Lavy, S., & Culp, C. H. (2012). Need for an embodied energy measurement protocol for buildings: A review paper. *Renewable and Sustainable Energy Reviews, 16*, 3730–3743.

Droege, P. (2006). *The renewable city*. Singapore: Wiley.

Downie, J., & Stubbs, W. (2011). Evaluation of Australian companies' Scope 3 greenhouse gas emissions assessments. *Journal of Cleaner Production,* In Press, Corrected Proof, pp. 1–8.

Gollagher, M., & Hartz-Karp, J. (2013). The role of deliberative collaborative governance in achieving sustainable cities. *Sustainability, 5*(2013), 2343–2366.

Hartz-Karp, J. (2012). Laying the groundwork for participatory budgeting—Developing a deliberative community and collaborative governance: Greater Grealdton. *Journal of Public Deliberation. 8*(2), 6.

Hernandez, P., & Kenny, P. (2009). From net energy to zero energy buildings: Defining life cycle zero energy buildings (LC-ZEB). *Energy and Buildings, 42*(6), 815–821.

Kennedy, S., & Sgouridis, S. (2011). Rigorous classification and carbon accounting principles for low and Zero Carbon Cities. *Energy Policy 39*, 5259–5268.

Newman, P., & Jennings, I. (2008). *Cities as sustainable ecosystems: Principles and practices.* Washington DC: Island Press.

Newman, P., & Kenworthy, J. (2015). *The End of Automobile Dependence: How Cities are Moving Beyond Car-based Planning.* Washington DC: Island Press.

Newton, P., & Newman, P. (2013). The geography of solar PV and a new low carbon urban transition theory. *Sustainability, 5*(6), 2537–2556.

Newton, P., Newman, P., Glackin, S., & Trubka, S. (2012). Greening the greyfields: Unlocking the development potential of middle suburbs in Australian cities. *World Academy of Science, Engineering and Technology, 71*, 138–157.

Reid, M. S. (2008). Stakeholder participation for environmental management: A literature review. *Biological Conservation, 141*, 2417–2431.

Thabrew, L., Wiek, A., & Ries, R. (2009). Environmental decision making in multi-stakeholder contexts: Applicability of life cycle thinking in development planning and implementation. *Journal of Cleaner Production, 17*, 67–76.

Williams, J. (2012). *Zero Carbon Homes: A Road Map*. New York: Earthscan.

World Business Council on Sustainable Development (WBCSD). (2009). *Transforming the market: Energy efficiency in buildings.* Retrieved from http://www.wbcsd.org/transformingthemarketeeb.aspx.

World Resources Institute & World Business Council for Sustainable Development. (WRI, & WBCSD) (2004). *The greenhouse gas protocol: A corporate accounting and reporting standard*-revised edition. Retrieved from http://www.ghgprotocol.org/files/ghgp/public/ghg-protocol-revised.pdf.

World Resources Institute & World Business Council for Sustainable Development. (WRI, & WBCSD) (2011). *Corporate value chain (Scope 3) accounting and reporting standard.* Retrieved from: http://www.ghgprotocol.org/standards/scope-3-standard:.

Yang, J., Shen, G. Q., Ho, M., Drew, D. S., & Xue, X. (2011). Stakeholder management in construction: An emperica; study to address research gaps in previous studies. *International Journal of Project Management, 29*, 900–910.

Zuo, J., Read, B., Pullen, S., & Shi, Q. (2012). Achieving carbon neutrality in commercial building developments—Perceptions of the construction industry. *Habitat International, 36*, 278–286.

Chapter 12
A New Framework and Core Elements

12.1 Introduction

From the issues outlined in previous chapters, it is clear that there are several obstacles currently facing the urban-planning/development sector, which make gaining various forms of carbon certification difficult. It is argued that new sector-specific guidelines for urban development are required in order to provide greater clarity and direction for developers wishing to understand their carbon footprint and pursue certification. It will also help to bring greater credibility to the concept of carbon neutrality for the built environment and allow accurate and trustworthy comparisons to be made. This is true not only for carbon neutrality, but for any type of carbon claim. The following section proposes a framework for calculating the emissions associated with precinct-scale low-carbon development, and it outlines some core elements that need to be considered when establishing a certification process for such development.

12.2 Proposed GHG Framework for Precinct-Scale Urban Development

The GHG framework for precinct-scale urban development proposed by the authors involves the calculation of emissions associated with six key areas of development. Each area has its own literature about the benefits of adopting alternative technologies or approaches, although the process of bringing them together has not been undertaken. Thus, the approach here is to develop a detailed understanding of the carbon associated with each of the six areas of an urban development project.

The areas have been chosen based on the review conducted in Chaps. 3 and 4, which identified significant carbon abatement opportunities, as well as other savings (such as in potable water use), if alternative practices are adopted.

V. Rauland and P. Newman, *Decarbonising Cities*, Green Energy and Technology, DOI 10.1007/978-3-319-15506-7_12

Some of the areas identified are often neglected by developers, as they commonly fall outside the operational control, and therefore responsibility, of a developer (e.g. energy and transport infrastructure). However, they have been chosen and included in the analysis for the reason that they can be influenced by the developer and they are of significant interest to planning agencies. Tools that include all of these characteristics can become important design tools for achieving planning approval. Furthermore, developers are now acknowledging not only the carbon reduction possible from transit-oriented development (Newman and Kenworthy 1999) but also the financial value in locating urban development around public transport systems (McIntosh et al. 2011).

New small-scale technologies and processes for supplying and managing resources, such as energy, water and waste at the precinct level, are now being identified as key priorities for developers and are increasingly targeted in the design of developments (Bartholomew and Ewing 2009; Newman et al. 2009; Newman and Matan 2013; Roseland 2012).

The six main sources of emissions, which this book proposes should be included in any carbon analysis of the built environment, include the emissions associated with the:

1. Site preparation and construction process
2. Embodied carbon in materials
3. Energy production and management
4. Water management
5. Waste management
6. Transport

These sources are illustrated in Fig. 12.1. Examples of low-carbon options available within each area are provided below.

12.2.1 Site Preparation and Construction Process

The first step in the proposed framework for calculating the emissions associated with precinct-scale urban development involves site preparation and the construction process.

Site preparation involves the demolition, management and disposal of existing, onsite building, construction and infrastructure materials (i.e. if development is occurring on grey- and brownfield sites), along with other processes involved in preparing a site for development, e.g. excavation, leveling and tree removal (Boake 2008).[1] The *construction process* involves the emissions associated with different

[1] Boake (2008) notes that significant emissions can also be locked up in the natural landscape (e.g. trees) before development, which he suggests should also be accounted for in carbon calculations.

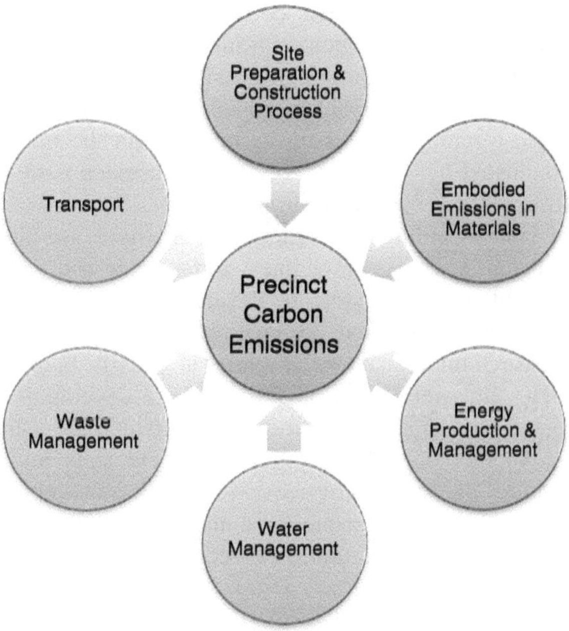

Fig. 12.1 Key emission sources within precinct development

methods for constructing buildings and infrastructure, for example, using prefabricated materials versus onsite construction. These two areas are discussed further below.

12.2.1.1 Site Preparation

With the above issues in mind, it is argued that demolition and disposal (D&D) emissions are best captured in the beginning of the lifecycle analysis and development process under the banner of site preparation. This would help to ensure that the emissions captured are based on actual numbers, rather than assumed or predicted in the future. It will also ensure that a developer takes part in managing onsite waste responsibly in order to minimise GHG emissions. In addition, it can help developers to become aware of the significant emissions already locked up in the existing materials (which they do not need to account for if reused) compared to the emissions embodied in creating or using new materials (which will need to be accounted for). This may help to incentivise developers to think more holistically about the reuse of materials and maintenance of existing infrastructure onsite—i.e. seeing waste as a resource, as opposed to razing a development to begin from scratch. This is important, as research shows that 'building reuse almost always yields fewer environmental impacts than new construction when comparing buildings of similar size and functionality' (Preservation Green Lab 2011: vi).

There may, of course, be situations where it is unfeasible to maintain existing buildings or infrastructure, for instance, in the case where low-density urban form is being transformed into high-density development, or where the land use of the site changes from industrial to residential. However, there may still be opportunities to reuse certain materials onsite or to retain the outer shells of buildings.[2] A study conducted by Thormark (2002) demonstrated a 45 % reduction in GHG emissions from using recycled materials.

Understanding that significant emissions are already locked in essential infrastructure such as power lines, water pipes and roads may also make infill development more attractive to developers, who will not have to account for these emissions, compared to greenfield development, where they would need to. Research undertaken by Trubka et al. (2010a, b, c) demonstrates not only additional GHG emissions associated with fringe development, but also hidden costs associated with this infrastructure—costs which are usually borne by the tax payer.

Another consideration for greenfield sites involves tress. If site preparation requires the removal of trees without adequate replacement, it is recommended that the carbon already sequestered and stored in them, as well as carbon that would be sequestered in the future, be accounted for in the calculation process, as Boake (2008) suggests.

Understanding and accounting for all the emissions associated with site preparation can influence decisions made by developers and can dramatically affect the carbon emissions associated with a development.

12.2.1.2 Construction Process

Emissions are generated during the construction process of developments through onsite power and water use and the transport of materials to site, as well as from waste produced from unused materials (i.e. construction and demolition (C&D) waste). Currently, around 42 % of waste generated in Australia comes from C&D and around 50 % of this ends up in landfill (Productivity Commission 2006). There are also emissions associated with transporting waste to landfill or a recycling facility.

However, the GHG emissions associated with these processes can vary considerably depending on the construction approach and techniques used. The use of prefabricated (pre-fab) materials,[3] for example, has many environmental and GHG benefits, such as the reduction of waste produced onsite compared to traditional construction approaches, fewer embodied emissions in the materials due to greater potential for onsite reuse of materials during the manufacturing process and a reduction in the time needed to construct a development, which in turn reduces the amount of onsite power (electricity) and transport fuel required (Aye et al. 2012;

---

[2] Examples of this are include old warehouses being converted to residential dwellings.

[3] This is a process whereby materials or components of a building or structure are assembled in a factory and then delivered as larger components or as complete assemblies to the construction site.

Lehmann 2013). All this contributes to reducing not only onsite GHGs, but also costs, which can increase the affordability of buildings, and in particular of housing (Lehmann 2013).

It should be noted that there is a significant overlap between site preparation, the construction process and the embodied emissions in materials, discussed in Sect. 12.2.2.

12.2.1.3 Dealing with Demolition and Disposal

While there appears to be limited academic analysis of site preparation as an isolated activity for GHG analysis within the development industry, much research has identified the importance of including the GHGs associated with building construction, demolition and disposal of waste in analyses, particularly if taking a life cycle approach to emission calculation (Crawford 2009; Dixit et al. 2012; Nassen et al. 2007; Thormark 2002). However, calculating the D&D emissions from individual buildings is far easier than calculating them from precinct-scale infrastructure at the end of its life (i.e. at some unknown time period in the future), which proves rather complicated and impractical for several reasons highlighted below.

- ***Differing Life Cycles of Infrastructure***: To begin with, precinct-scale development involves the construction of different types of infrastructure, such as pipes, roads, houses, buildings and energy infrastructure, all which have differing life expectancies and will thus need to be upgraded at different times. This makes traditional life cycle analysis of GHG complicated for a precinct-wide development.
- ***Uncertainty Around Demolition and Disposal Processes***: It is also difficult to predict what processes will be employed when demolishing parts of a precinct development at some future point in time, and how the waste will be disposed of (i.e. if it will be sent to landfill, incinerated, reused or recycled). The calculation of these emissions will be based on significant assumptions, which may or may not ultimately hold true.
- ***Double Counting***: Any processes or activities undertaken by developers in preparing their land for construction, such as demolishing existing infrastructure and disposing of the waste, logically become their responsibility and should be taken into account in their GHG analysis. However, if D&D emissions have been included in an LCA of the previous developer, it would mean these emissions do not need to be accounted for by the current one, despite them undertaking the actual process. If this is the case, not only would there be little incentive to dispose of waste responsibly (if it means additional cost or effort), but this situation could also lead to issues of double counting, i.e. if emissions are calculated by a developer at the beginning and by another developer at the end.
- ***Redundancy***: Factoring in the D&D emissions from the start could arguably be seen as building in redundancy, rather than assuming that materials and infrastructure will last or be adapted or reused.

12.2.2 Embodied Carbon in Materials

As mentioned in Chap. 3, embodied energy (EmE) typically refers to the emissions associated with the production and transportation of materials used to construct buildings (Cabeza et al. 2013). While the emissions associated with onsite construction, D&D are also often captured under the banner of EmE from a life cycle perspective; the definition proposed in this book confines it just to the emissions associated with the extraction of raw materials, processing, manufacturing and transport, also known as 'cradle-to-gate' emissions (Dixit et al. 2010). The emissions associated with the construction process (including how materials are transported to and moved around on the site), as well as emissions from demolition, are captured separately in this analysis (i.e. in the above-mentioned construction process).

Recent research shows that EmE is responsible for a greater proportion of a building's life cycle emission than previously thought and can be as high as 62 %, largely because increasing energy efficiency within buildings reduces operational emissions by comparison (Cabeza et al. 2013; Crawford 2009; Sturgis and Roberts 2010; Thormark 2006).

EmE is thus included as a key component in the carbon analysis of precinct-scale development, not only due to the substantial amount of emissions associated with EmE, but also because of the significant emissions reduction possible depending on the materials used. For example, choosing to reuse materials that are already onsite or purchasing recycled materials generally produces far fewer emissions than using new materials (Preservation Green Lab 2011; Thormark 2002), depending on the type of material and its production process.

Many new low-carbon materials are also being developed, materials that use either low-carbon supplements, such as fly ash in cement manufacturing, or cleaner production processes such as the use of renewable energy (Cabeza et al. 2013). Some studies have shown that the amount of embodied water in materials could be just as significant as the amount of energy (Crawford and Treloar 2005), which can also have a significant impact on emissions, particularly if the water used is sourced from energy intensive processes such as desalination.

Finally, EmE can also vary considerably depending on where materials are sourced, i.e. locally, regionally or internationally, and how they were transported to site.[4]

[4] When undertaking carbon accounting at the precinct level, it is important to clarify where emissions have been captured (e.g. emissions from the transport of materials used in construction could be captured under either embodied emissions or the construction process). This is important to clarify in order to avoid double counting or not counting at all.

12.2.3 Energy Production and Management

Energy is a particularly important component of the carbon framework, as the majority of life cycle emissions from the built environment generally come from energy-related activities (Newton and Tucker 2010). However, the emissions associated with energy production and consumption within a development can vary depending on a variety of factors such as the design and type of buildings, thermal efficiency, appliance efficiency, occupant behaviour and renewable energy generation (Newton and Meyer 2012; Newton and Tucker 2010).

Different building types have been examined in terms of their energy efficiency, with studies showing that denser urban form[5] is generally more thermally efficient than single detached housing (Glaeser and Kahn 2008). The design of buildings can also dramatically affect the amount of energy required to run them. Well-known building techniques such as solar passive design have been extensively studied and are shown to significantly reduce the amount of artificial heating and cooling required in buildings (Morrissey et al. 2011; Stevanović 2013). There are various energy efficiency measures such as improved insulation (Jelle 2011), and alternative heating and cooling technologies that can also significantly reduce energy requirements (Finney et al. 2012).

Significant variations in emissions can also occur depending on the source of the electricity supplied to buildings within a development. As noted in Chap. 4, conventional grid electricity sourced predominantly from coal is the most emission intensive form of electricity production and the primary cause of climate change (IPCC 2007). Electricity sourced from natural gas produces fewer emissions and is particularly efficient when used in a way that supplies power and heating (co-generation) or power, heating and cooling (tri-generation). The notion of small-scale, decentralised energy production (both from fossil fuel and renewable sources), is becoming increasingly popular as a low-carbon source of power and is well suited to the precinct scale (Rauland and Newman 2011). Other decentralised low-carbon or carbon-free sources of energy include PV panels, solar hot water heating, small-scale geothermal heat pumps and small-scale wind.

Purchasing 'green power'[6] is another option, which in this case eliminates the emissions associated with electricity consumption onsite.

12.2.4 Water Management

The supply and treatment of water is another critical element to include in the GHG analysis of new developments. Water management is becoming increasingly

[5] For example, townhouses and apartments.

[6] GreenPower is a trade-marked name in Australia and refers energy that comes from 100 % renewable or carbon-free sources.

connected to energy and carbon emissions as more cities, particularly in Australia, begin to rely on energy intensive forms of water supply and treatment such as desalination. The connection between energy and water, and thus emissions, has been extensively examined in the literature (see Griffiths-Sattenspiel and Wilson 2009; Hall et al. 2009; Kenway et al. 2008; Medeazza and Moreau 2007; PMSEIC 2010; Rothhausen and Conway 2011; Siddiqi and Anandon 2011).

While GHG emissions can vary considerably depending on the technology used to produce water (e.g. through desalination, treating recycled water or using dam water catchments), emissions occur over the water entire cycle, including through the distribution of water (i.e. pumping of water around cities), the consumption of water and the treatment of wastewater (Wang 2009). While some studies have examined the environmental savings associated with decentralised water management (Sharma et al. 2010) and warn of the danger of locking into new energy intensive centralised systems such as desalination (Chanan et al. 2009), opinions still vary as to which is the most efficient or emission intensive form small-scale or large-scale (Fagan et al. 2010).

Regardless of the scale debates, research is showing that water has become one of the biggest energy users in cities (Kenway et al. 2008), and this is thus a critical area to tackle in urban development. Knights et al. (2007) question whether utilising renewable energy could be a way to offset the GHG impact of the rising energy use, though they warn of the issues associated with making offset claims.

The amount of water needed within a development will also vary depending on the urban design, which includes how gardens and public landscaping are managed. Water efficiency measures in buildings will affect the amount of water required as well (Kenway et al. 2008). Ensuring that water efficient appliances and grey water/third pipe systems are built into developments from the outset can optimise the water efficiency associated with the building infrastructure.

Development approaches that use water sensitive urban design (WSUD) require much less water for irrigation and can provide opportunities for alternative storm water treatment, all of which can dramatically reduce GHG emissions (Sharma et al. 2008). Optimising local water sources and recycling have many environmental benefits (Hatt et al. 2005; Sharma et al. 2010) and can dramatically reduce emissions, though few precinct-scale initiatives have assessed this.

12.2.5 Waste Management

As mentioned in Chap. 4, the waste sector in Australia produces around 11 MT of GHG emissions annually (DSEWPC 2010). The GHG emissions associated with managing solid waste can vary considerably depending on the waste management technique adopted. Currently, around half of the waste produced in Australia ends up in landfill, which is a particularly inefficient and emission intensive form of waste management due to the production of methane emissions during the anaerobic digestion of organic materials (Lou and Nair 2009).

The energy-related GHG emissions associated with solid waste management and different treatment strategies have been reviewed in academic literature such as Hughes et al. (2006) and Chen and Lin (2008). Waste management options that produce fewer emissions include waste to energy technologies through incineration and various forms of gasification (Mountouris et al. 2006).

However, the concept of resource recovery within waste management is the optimal way of dealing with waste, that is, viewing waste as a resource (Tillman Lyle 2004), and this will no doubt play an essential role in decarbonising cities into the future. Resource recovery essentially involves reusing, recycling and composting as much of the waste as possible (Seadon 2010). This not only has the potential to significantly reduce emissions, but also helps to create a more circular metabolism within our cities as espoused by Wolman (1965) and Giradet (2004).

The very first step in any waste management process, however, should be to minimise or avoid the amount of waste generated in the first instance. Encouraging society to minimise waste is likely to be challenging and will require local behaviour change programs. Increasing knowledge around recycling can also be facilitated by ensuring sufficient infrastructure is in place to enable good behaviour (e.g. clearly marked recycling points). New waste collection techniques such as municipal vacuum waste, which sucks waste through an underground network of pipes to one or several collection points within a city, can also help to encourage better recycling. It can also dramatically reduce the transport emissions associated with waste collection, improve the efficiency of recycling and add amenity by reducing the number of garbage trucks on the road (City of Sydney 2010). This system can also allow for more compact urban form by eliminating the need for roads large enough for garbage trucks to manoeuvre in (a traditional design factor for cities). It is argued that ongoing waste management therefore be considered as part of precinct design, in additional to the construction-based waste management.

12.2.6 Travel

Transport fuels used by residents will impact the GHG emissions associated with an urban development. Different modes of transport (e.g. trains, light rail, buses, traditional cars and electric vehicles) generate varying amounts of GHG emissions. Vast academic literature has analysed the differences in transport GHG emissions associated with different urban forms both internationally (Glaeser 2011; Newman and Kenworthy 1999, 2015), and in the Australian context (Mees 2010; Newman 2006; Trubka et al. 2010a, b, c), with a strong correlation being demonstrated between higher density and lower transport-related GHG emissions (Newman and Kenworthy 1999, 2015).

Moreover, even between cities with the same mode of transport (e.g. trains), emissions can vary depending on the fuel source. For example, the emission intensity of an electricity grid will affect transport emissions associated with rail, if it is drawing

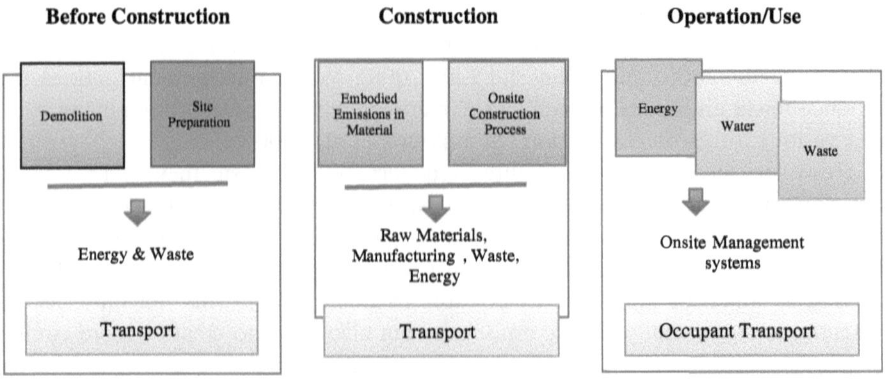

Fig. 12.2 Phases of precinct-scale urban development emissions

electricity from the grid. Electric-based transport systems (i.e. trains, light rail or electric vehicles) that recharge using renewable energy can significantly reduce transport-related emissions (Newman et al. 2012; Newman and Kenworthy 2015).

Urban development that encourages low-carbon and carbon-free transport modes, such as walking, cycling and access to efficient public transport, can dramatically reduce the amount of GHG emissions associated with a development. It is therefore essential that these emissions be factored in when designing a development, and then regularly monitored.

12.2.7 Phases of Precinct-Scale Emissions

Figure 12.2 illustrates the various stages of development and the proposed emissions that should be captured within carbon analysis.

12.3 Core Elements for Low-Carbon/Carbon Neutral Certification Within the Built Environment

12.3.1 A Carbon Accounting Framework for Urban Development

The previous section outlined the six key emission sources that are proposed as the broad framework for calculating precinct-scale urban development emissions. The following section provides some insight into and recommendations for undertaking the carbon footprint analysis at the precinct level.

12.3.1.1 A Hybrid Approach

The carbon measurement process and methodology for most applicants seeking carbon neutrality is relatively clear and straightforward: either they adopt a GHG inventory approach or a life cycle analysis approach. However, the National Carbon Offset Standard's (NCOS) CNP guidelines acknowledges that there are some areas, such as construction projects, that 'may not obviously fit into these categories' (DCCEE 2011: 6). Suggesting that a hybrid methodology may be required in some cases, the guidelines recommend that such projects should follow up with a program administrator to determine the appropriate process for their specific circumstances (DCCEE 2011). This has significant implications in terms of the replicability of, and comparability between, development projects if no underlying policy or framework exists to guide administrators in the advice they provide applicants. Different administrators with varying levels of knowledge and expertise may provide applicants with very different advice, which would then produce very different carbon outcomes.

It is therefore suggested that the framework provided in this chapter is to be adopted as the basis for the carbon footprint for developers wishing to claim carbon neutrality or any carbon claim (e.g. low carbon). The general process required for calculating the emissions would still need to follow the standard carbon accounting rules outlined in the existing standards and protocols.

12.3.1.2 Dealing with Existing Development/Materials

To avoid double counting sites where infrastructure already exists, it is recommended that developers do not include the carbon associated with this, as these emissions have already been locked up, and it is assumed that the previous developer will have accounted for these emissions (e.g. in the future, when emissions reporting is more mainstream). This should make developing on existing land more desirable to developers, as fewer emissions would need to be accounted for, than developing on greenfield sites.

12.3.2 Online Modelling Tools

12.3.2.1 New Tools to Assist with the Calculation Process

The carbon accounting process is likely to be seen as too complicated and time consuming for many developers attempting to undertake the analysis on their own, as the process requires in-depth knowledge of carbon accounting. It is therefore recommended that new carbon calculation tools be developed urgently to assist the development industry to undertake this analysis, which will also help to add consistency in the measurement process.

The development of twenty-first century design tools that can be easily and affordably accessed by developers, the community and planning agencies are likely to become a major part of the urban development process over the next few years (Newton et al. 2012). They will be an important part of the decarbonisation process for urban development as well as the certification process.

Having access to new, easy-to-use and intuitive online carbon footprinting tools for development projects will make the process of calculating precinct emissions far easier and more cost-effectiveness for developers.

As mentioned in Chap. 9, eTool and Precinx are two examples of existing online tools that focus on the carbon associated with the built environment (as well as other important factors in the assessment process) and go some way in assessing these emissions. They both will require additional elements in order to incorporate the six areas identified in this chapter.

While eTool is currently free to use (though fees are required for consultancy assistance, and to verify and certify emissions), the costs associated with using the PRECINX tool are currently unknown, as the tool is only being used by developers and state government land development agencies and is not publicly available. The developer of this tool, Kinesis, also has another version of the tool (C^{CAP} Precinct), which developers can licence for a fee based on the size of a project. Further development of a publicly available online tool for use in design and assessment is now underway as part of several research projects funded by the Federal Government.[7]

For such tools to be used in a certification process, such as the NCOS CNP, a key requirement is transparency. As discussed in Chap. 11, all carbon audits need to be verified by a third party in order to gain certification. This would mean that a third party (and the certifying body) would need to be able to see and verify carbon calculations and assumptions sitting behind such tools. This is a currently an issue with existing tools, as their owners are justifiably unwilling to expose the calculations underpinning their software for intellectual property (IP) and other confidentiality reasons.

If a standard modelling tool can be developed for precinct-scale design and assessment, it will break this nexus and is likely to become the standard required by all urban development at precinct scale. The other tools could still be used if they added value to what the standard model delivered. However, the results would still need to be verified by a third party, and the certifying bodies would need to be able to view the calculations that sit behind the tools in order for them to become approved by the appropriate bodies. The commercially sensitive information sitting behind the tool could still remain confidential while ensuring that the basic carbon analysis was being acceptably conducted.

It is expected that as demand increases for such development-scale carbon measurement processes, new tools will become available, increasing market

[7] See http://crcsi.com.au/research/4-5-urban-planning/.

competitiveness and reducing the costs of using them. The move to design and assessment online will be sure to dramatically reduce the costs of all planning assessment. Hopefully, carbon assessment will just become another part of that process.

12.3.2.2 Modelling the Cost of Abatement

Another advantage associated with emerging online measurement tools is their ability to measure not only the carbon abatement associated with alternative designs, technologies and efficiency measures, but also the differences in costs.

While there has long been a wealth of information available regarding various carbon mitigation and adaptation actions and measures, there has been little systematic way of analysing and comparing those options in order to determine which are the most beneficial, appropriate and financially viable options in a particular situation or for a particular site.

Research and practice have highlighted the previous lack of information and tools available to support the process of choosing the most appropriate set of actions to meet a specified carbon reduction goal. Carlisle and Bush (2009) refer to this as 'the Planning Gap'. Gore (2009) acknowledges that 'cities and towns could benefit by developing computerised statistics on each of the major challenges they face and integrate them and display them visually for groups that include department heads and other stakeholders in a shared effort to discover what really works and what does not' (Gore 2009).

By enabling users to instantaneously see the impact each abatement option has on the overall carbon and costs associated with a development, these tools can help inform decisions to enable the optimum choices to be made. Without these new and innovative evidence-based tools, such as eTool and PRECINX, decisions will continue to be limited to check lists, which have proliferated globally and dominate current decision-making.

12.3.2.3 Tools to Reduce Certification Costs

Transaction costs may pose a significant threat to developers wishing to pursue carbon neutral certification for their development. As is commonly the case with built environment certification schemes, such as the Green Building Council of Australia's Green Star certification program and low-carbon Australia's NCOS Carbon Neutral Program, administration fees (including third party consulting fees) are a relatively large barrier and deterrent for many seeking certification.

However, as seen in Chap. 11, the certification fees paid to low-carbon Australia for their role in administering the process are not overly inhibitive, particularly when compared to the entire budget associated with developing land. The largest cost is expected to be the work associated with the initial and ongoing calculation of a precinct development's life cycle emissions/carbon footprint, and this is likely to pose the greatest challenge in terms of time and cost burden on a developer and/or the body responsible for pursuing the ongoing carbon neutral status. However, the development of new online tools, as discussed above, can help to overcome this barrier.

12.3.3 Governance—Ongoing Management of Low-Carbon Precincts

12.3.3.1 Management of Emissions and Annual Reporting

While the initial carbon footprint of a development can be managed by the developer (assuming they have the appropriate in-house knowledge, tools and/or assistance from a carbon accounting specialist), the ongoing, annual reporting of emissions and the development of emission management plans (EMPs) are two significant issues for developers who wish to maintain their development's carbon neutral or other carbon status beyond their own involvement in the construction. As mentioned in Chap. 11, developers usually walk away from any responsibility for a development once it is complete and has been sold. The primary issue, therefore, is that of who will have the ongoing responsibility for measuring the annual GHG emissions associated with developments and maintaining certification. This is discussed under governance in the section below.

12.3.3.2 Developing an Emission Management Plan

Writing up an annual emission management plan (EMP), which identifies continual abatement measures for a precinct-scale development, will be quite different to traditional EMPs as the development is likely be designed as highly carbon efficient from the beginning. For example, compared to organisations and businesses that often benefit from having abundant retrofit and energy efficiency opportunities to pursue, well-designed precinct-scale development should have few carbon abatement opportunities. Nevertheless, there will always be chances to upgrade elements of developments when new technologies become available. A particular area for ongoing management will be around occupant engagement and education.

12.3.3.3 Governance Structures

This book proposes that a governance structure needs to be established or a body identified or formed to undertake, manage and be held responsible for the process of measuring emissions and maintaining certification of ongoing operations in the built environment sector. Strata companies and local governments are identified as being in the best position to take on this responsibility. Green leases are also identified as a mechanism to assist the ongoing management of the precinct. These are briefly discussed below.

12.3.3.4 Strata Schemes, Companies and Body Corporate

Strata schemes are group residential and/or commercial building arrangements, which allow individual ownership of a strata lot (a single premise within a group of premises) and community ownership and management over shared facilities such as gardens, lifts and swimming pools (Western Australian Land Information Authority 2012). Due to the shared facilities, the cost of buying into a strata scheme is usually more affordable than buying freehold title properties. They can also reduce the costs associated with shared infrastructure services such as electricity, gas, sewerage and water and minimise the maintenance costs associated with buildings (e.g. painting).

All strata schemes have a Body Corporate or Strata Company that manages the day-to-day operations and maintenance of shared facilities and makes the decisions, which affect all owners. Every owner on a Strata-Plan is automatically a member of the Strata Company.[8] This company, or elected Strata Council,[9] also formulate by-laws[10] specific to their complex (Western Australian Land Information Authority 2012).

A strata company could be an appropriate body to administer and maintain responsibility for a development's carbon neutrality status, as it already has a governance framework that would enable this sort of administration to be carried out. Having a history of administering levies and rates and being responsible for shared infrastructure, body corporates, through their strata companies, could set up a system for requiring certain fees to be paid annually (e.g. for consultant costs for carbon footprinting, purchase of offsets and payment of CN certification fees).

Strata companies can also create by-laws that require owners within the complex to give the body corporate access to their utility bills, which would be needed for the carbon footprint at the end of each year. However, submetering to the site (i.e. for electricity, gas and water) is likely to already be in place within a strata complex, which may make such a by-law unnecessary.

Further, by-laws may require new owners and tenants to undertake a compulsory sustainability or low-carbon training or induction session (discussed further below) and to agree to live by certain sustainability principles. It may also require residents to undertake a bi-annual survey of their transport modes and travel emissions (also required for the precinct's carbon footprint measurement), as well as informing the body corporate of any individual infrastructure upgrades or renovations undertaken within their apartment/space, in order to capture all onsite embodied emissions. This may also already be required as a by-law for other reasons.

---

[8] In the case of Western Australia.

[9] A Strata Council is created in situations where owners do not wish or do not have time to partake in everyday decision-making regarding their complex.

[10] A by-law is a rule by which strata owners must abide.

Compulsory green leases could be mandated and managed by strata companies or local councils to ensure properties comply. These are discussed after the following section.

12.3.3.5 Local Government

Local councils are another entity that would be suitably placed to manage the ongoing carbon measurement process and maintenance of certification for carbon neutrality. Many councils have already been measuring GHG emissions at the local government level as part of their commitment under ICLEI (see Chap. 8), and so have experience in dealing with GHG measurement. There are also online tools available for calculating and reporting operational emissions at the local government level in Australia.[11]

Local governments also have a history of administering and receiving rates and levies from landholders within their jurisdiction. A carbon neutral levy could thus be built into the existing rates system but applied specifically to low-carbon/carbon neutral districts or neighbourhoods. This carbon neutral levy would capture the costs associated with undertaking the annual GHG measurement process (as specified under the NCOS CNP guidelines) and EMP for the precinct and, if desired, the purchase of offsets and carbon neutral certification fees.

This levy could also potentially facilitate the implementation of low-carbon infrastructure, such as precinct-scale tri-generation, by enabling the upfront costs associated with installing the system to be spread out over a longer time frame and repaid through local government rates. This would help to reduce a developer's upfront costs, which would be passed through to the buyer, thus increasing the affordability of buying into the area. While land rates may be higher than other areas, this will be offset by reduced operation costs (i.e. energy and water bills). The concept (a GRID), is discussed further in Chap. 13.

Local councils have a vested interest in ensuring that the optimal precinct-scale solutions are reached, as this will not only reduce the carbon associated with their local government area (LGA),[12] but also have the potential to increase the long-term affordability of living and working there,[13] and thus the attractiveness of locating business and residents there.

Behaviour change is another issue that could also be addressed at the local government level. For example, local councils could provide free sustainable living courses to residents moving into the area,[14] though it would be difficult for a

[11] See the WALGA Reporting Platform—http://www.walgaclimatechange.com.au/walga-reporting-platform.htm.

[12] Which is likely to increase in importance in coming years.

[13] For example, by reducing the utility costs associated with energy, water and waste management.

[14] Several councils in Western Australia already offer this through the Living Smart program.

council to mandate such a program. Behaviour change is discussed further in Sect. 12.3.4 below.

12.3.3.6 Green Leases

A Green lease is a contract between the tenant and the building owner that sets out rules and guidelines for optimising the environmental performance of buildings (e.g. for improving energy and water efficiency and waste management within the buildings). The lease is negotiated between the two parties to ensure that the measures, actions and responsibilities for each party are realistic, manageable and implementable. Traditionally, green leases have been applied to commercial properties as a mechanism to help overcome the problem of split incentives discussed in Chap. 11 (and discussed further in Chap. 13).

Green leases allow for a more collaborative approach to dealing with upgrades and general operational performance of buildings and can create a more amiable relationship between landlord and tenant (COAG 2012). Green leases often provide an opportunity to create an Environmental Management Committee (EMC), which, like other important stakeholders such as building and facility managers, enables landlords and tenants to meet regularly and to develop and review annual Environmental Management Plans (Christensen and Duncan 2007).

It is recommended that green leases be used when subletting or leasing property within a carbon neutral development or precinct. Strata by-laws may already provide some mandatory environmental efficiency requirements for owners. Green leases could therefore provide a suitable mechanism to ensure that tenants moving into the area also accept responsibility for ensuring that the environmental standards and goals of the precinct are met.

12.3.4 Engagement and Behaviour Change Programs for Operational GHG

Occupant behaviour remains a key issue in achieving the sustainability and low-carbon outcomes associated within precinct-scale development (Zuo et al. 2012). The design of low-carbon precincts, incorporating various types of low-carbon infrastructure, is essential in facilitating and enabling a low-carbon lifestyle for residents and businesses living and working in the area. Fuller (2009) highlights the importance of designing spaces and environments 'to improve outcomes by influencing patterns of choice' (15). Once the design has been certified, the next stage is to ensure that the owners and occupiers of the precinct are able to keep within the expectations of the certification, or to even go further.

Having the infrastructure in place does not necessarily mean that people will adopt the behaviour associated with a low-carbon lifestyle, however. Much literature has documented the problems associated with basing behaviour change strategies for sustainability on the assumption that humans behave rationally (Fuller 2009; Jackson 2005; Moloney et al. 2010).

BedZED, one of the case studies discussed in Chap. 6, provides an apt example of the effect behaviour can have on the performance of a development. Based on the infrastructure available and the expected behaviour of 'ideal' residents, the per capita ecological footprint of the development was predicted to be 1.9 ha. However, after evaluating the actual behaviour of the residents, the average ecological footprint was shown to be 4.36 ha (Newman and Jennings 2008). It is important to note that this was not due to the infrastructure or development not performing properly, but more to do with people's attitudes and behaviour. For example, despite the availability of a car shares scheme and local public transport, some people still decided to own and drive private vehicles solely.

In order to create the most effective strategies to help transition society towards low-carbon communities, Moloney et al. (2010) suggest basing them on a 'socio-technical framework that considers both individual psychological factors as well as the systems, standards and norms under which individuals operate' (7614).

Fuller (2009: 3) suggests policies need to be created that provide opportunities to create 'choice contexts' for encouraging optimal social outcomes. He highlights five tools that can influence the choices people make. These are given as follows:

1. Setting the default option in a set of choices to the favoured option;
2. Offering 'self-contracting' to support commitment;
3. Presenting and organising information in a user-friendly way;
4. Designing physical spaces to guide behaviour; and
5. Supporting the development of social norms.

Many of these 'tools' can be applied to various aspects of a low-carbon or carbon neutral precinct development. Two options that go some way to addressing these issues are behaviour change programs such as Living Smart and Travel Smart (Ashton-Graham and Newman 2013) and the adoption of green leases, discussed above. This book suggests that these programs (or variations of them) should be considered as part of the requirements for moving into, and living, within a designated low-carbon precinct.

Further research into other strategies and mechanisms for influencing behaviour within low-carbon precincts is recommended and is currently being pursued through a Cooperative Research Centre (CRC) on low-carbon living.[15]

12.3.4.1 Living Smart and Travel Smart

The Living Smart program is a sustainability behaviour change program developed in the port city of Fremantle, Western Australia. It was the outcome of a call to

---

[15] See http://lowcarbonlivingcrc.com.au/research/program-3-engaged-communities.

action by local residents keen to learn more about sustainability. The program was developed by Murdoch University, the Fremantle Council and The Meeting Place Community Centre. It is a hands-on, interactive program that runs over 6–8 weeks and discusses a variety of sustainability issues that can be tackled in the home. The course encourages participants to set their own sustainability goals and get involved in weekly activities. Living smart is now an award-winning program with courses being organised and run all over Australia and has just begun in the USA (Newman and Kenworthy 2015).

Travel Smart is a similar sustainability-based program targeting travel behaviour. This program also originally began in Perth, Western Australia, and was so successful that it has now been adopted throughout Australia and internationally. Unlike the Living Smart program, Travel Smart is conducted mainly over the phone or may involve someone coming to a program participant's house to discuss travel options and provide information about local public transport services or other travel options (e.g. bike paths). The aim was to empower people to make better, more sustainable and healthier travel choices by giving them the right information. Evaluations show significant savings in GHG are achieved by these programs (Ashton-Graham and Newman 2013).

These two courses, or similar ones addressing behaviour change around the world, could be built into a carbon neutral management plan for local residents and businesses within the precinct. Even if it is not compulsory or undertaken by all, if enough people undertake it and demonstrate the desired sustainable behaviour, this can help to create a social norm, which can affect others' choices as people strive to conform (Fuller 2009).

12.3.5 Regulations to Assist Certification

New regulations will be required by government planning agencies that are ultimately responsible for land use, in order to establish a certification process that recognises the carbon reduction from urban development using the proposed framework and addressing the items identified as core elements. This body is likely to be the state government within the Australian system. These regulations would also contain mechanisms for resolving the question of offsets and the role of onsite renewables for those developers seeking carbon neutrality. These are discussed further below.

12.3.5.1 New Offsetting Rules Needed for Development Industry

As identified in previous chapters, a change in government in Australia and the introduction of new carbon policies has meant a range of activities from various sectors are now eligible to generate offsets, which can be used within the Australian NCOS CNP. This means that energy-related carbon abatement delivered onsite within precinct-scale developments in Australia can now potentially be used as an offset against other onsite emissions. This will have a significant effect on reducing the costs of pursuing and retaining carbon neutral status, and increases the incentive for developers to become net positive.

It is recommended that only activities that can generate carbon abatement above what is operationally required by a development (i.e. BAU), such as excess renewable energy generation, be considered as an eligible offset that could be used against other unavoidable emissions associated with developments, such as the embodied emissions in the materials used in construction. There is also the potential for developments to pursue offset generating activities within the bioregion that meet the requirements set out under the CFI/ERF.

If a developer does not generate excess renewable energy onsite, or cannot produce offsets through other CFI certified activities, then traditional offsets will need to be purchased (e.g. from those specified under the NCOS CNP) in order to neutralise the development's life cycle emissions. This can be done in several ways as discussed below.

12.3.5.2 Lump Sum Offsetting by the Developer

How developers go about offsetting a development's overall emissions may vary. For instance, as soon as a developer has measured and determined the total emissions associated with their development (including site preparation, construction process and EmE), they may choose to purchase a lump sum of offsets sufficient to neutralise the entire embodied emissions so that only the operational emission need to be accounted for on an annual basis.[16] They could then integrate the costs of this purchase into the individual land sales of the development, which, it is anticipated, would benefit from increased marketability.

This option assumes that, even if excess renewable energy is generated onsite, it will not be used to offset the embodied emissions associated with the development. The excess renewable energy could, however, be used to offset other ongoing operational emissions associated with the development, such as residents' transport emissions or emissions associated with the use of natural gas.

12.3.5.3 Offsetting Over an Extended Period

Alternatively, the amount of offsets required to neutralise the embodied emissions could be spread over the lifetime of the development.[17] If purchasing offsets, this would mean paying for the required offsets on an annual basis, potentially through

---

[16] The embodied emissions associated with renovations and upgrades will still need to be accounted for as they occur.

[17] However, this may be more difficult to calculate, considering the varying lifetimes of different types of infrastructure.

strata fees or council land rates as mentioned above. Or if a development is generating offsets (i.e. ACCU's) from activities onsite, these could be retired on an annual basis to offset embodied emissions.

12.4 Conclusion

This chapter has proposed a framework for calculating the emissions associated with precinct-scale urban development, as the research conducted for this book has revealed that no such framework currently exists. The framework has identified six key sources of emissions that should be included in any carbon analysis of urban development. These were based on the findings presented in Chaps. 3 and 4 that revealed significant emissions reduction opportunities in each of the six areas, when alternative approaches were adopted. While carbon neutrality is discussed in some depth within this chapter, the proposed framework is designed for all developments seeking some form of recognition for carbon reduction.

While Chap. 11 revealed that Australia currently has one of the most well-designed and rigorous Government-led carbon neutral certification schemes worldwide, the generic guidelines highlighted many issues for developers pursuing certification. These issues were addressed in this chapter and presented as recommendations designed to assist certification schemes in developing sector-specific guidelines for the urban land development industry. As this sector is very different from general businesses and organisations seeking carbon neutrality, it will no doubt require a different set of rules and guidelines. The proposed recommendations aim to provide greater clarity and direction for developers pursuing certification, and to allow more accurate and trustworthy comparisons of carbon emissions to be made.

The significance of the proposed framework lies not just in its provision of a basis for carbon neutral certification: this framework has the potential to set an international, and thus universal, standard in carbon accounting for urban development. It will need to be supported by other elements including:

1. Ways of bringing multi-stakeholders together in organisations such as Strata Title Companies to ensure on-going operational GHGs are managed and improved continuously;
2. New regulations to allow on-site renewables to be used as offsets against other emissions to ensure this rapidly growing green urban development approach is facilitated;
3. Behaviour change programs as part of any green assessment process; and
4. Minimisation of costs of certification by ensuring that online design and assessment can be done using publicly available models that contain all the above framework on carbon emissions as well as other planning outcomes.

References

Ashton-Graham, C., & Newman, P. (2013). Case study: Living smart in Australian households: Sustainability coaching as an effective large-scale behaviour change strategy. In S. Fudge & M. Peters (Eds.), *International approaches to behaviour change: The global challenge to encouraging sustainable lifestyles*. London: Edward Elgar.

Aye, L., Ngo, T., Crawford, R. H., Gammampila, R., & Mendis, P. (2012). Life cycle greenhouse gas emissions and energy analysis of prefabricated reusable building modules. *Energy and Buildings, 47*, 159–168.

Bartholomew, K., & Ewing, R. (2009). Land use-transportation scenario planning: A meta-analysis. *Journal of the American Planning Association, 75*(1), 1–15.

Boake, T. M. (2008). The leap to zero carbon and zero emissions: Understanding how to go beyond existing sustainable design protocols. *Journal of Green Building, 3*(4), 64–77.

Cabeza, L. F., Barreneche, C., Miro, L., Morera, J., Bartoli, E., & Fernandez, A. I. (2013). Low carbon and low embodied energy materials in buildings: A review. *Renewable and Sustainable Energy Reviews, 23*, 536–542.

Carlisle, N., & Bush, B. (2009). Closing the planning gap: Moving to renewable communities. In P. Dröge (Ed.), *100 per cent renewable—Energy autonomy in action* (pp. 263–288). London: Earthscan.

Chanan, A., Kanasamy, S., Vigneswaran, S., & Sharma, D. (2009). A gradualist approach to address Australia's urban water challenge. *Desalination, 249*(3), 1012–1016.

Chen, T. C., & Lin, C. F. (2008). Greenhouse gases emissions from waste management practices using life cycle inventory model. *Journal of Hazardous Materials, 155*, 23–31.

Christensen, S. A., & Duncan, W. D. (2007). Green leases: A new era in landlord and tenant co-operation? *Australian Property Law Journal, 15*(1), 54–65.

City of Sydney. (2010). *Green infrastructure for Sydney*. Retrieved from http://www.sydney2030.com.au/development-in-2030/city-wide-projects/powering-sydney-allan-jones.

Council of Australian Governments (COAG). (2012). *The green lease handbook. National strategy on energy efficiency*. Retrieved from http://www.gbca.org.au/gbc_scripts/js/tiny_mce/plugins/filemanager/Green-Lease-Handbook-20120907-PDF.pdf.

Crawford, R. H. (2009). Life Cycle energy and greenhouse emissions of building construction assemblies: Developing a decision-support tool for building designers. In *Sixth Australian Life Cycle Assessment Conference: Sustainability Tools for a New Climate, February, Melbourne, Vic.*

Crawford, R., & Treloar, G. (2005). An assessment of the energy and water embodied in commercial building construction. In *4th Australian Life Cycle Assessment Conference, Sydney, 2005*. Retrieved from http://www.deakin.edu.au/dro/view/DU:30005640.

Department of Climate Change and Energy Efficiency (DCCEE). (2011). *Carbon neutral program guidelines: National carbon offset standard*. Version 2.0. Retrieved from http://www.climatechange.gov.au/sites/climatechange/files/documents/03_2013/cnp-guidelines.pdf.

Department of Sustainability, Environment, Water, Population and Communities (DSEWPC). (2010). *Reducing the materials and resource intensity of the built form in the perth and peel regions*. Canberra: Commonwealth of Australia.

Dixit, M. K., Fernandez-Solis, J. L., Lavy, S., & Culp, C. H. (2010). Identification of parameters for embodied energy measurement: A literature review. *Energy and Buildings, 42*, 1238–1247.

Dixit, M. K., Fernández-Solís, J. L., Lavy, S., & Culp, C. H. (2012). Need for an embodied energy measurement protocol for buildings: A review paper. *Renewable and Sustainable Energy Reviews, 16*, 3730–3743.

Fagan, J. E., Reuter, M. A., & Langford, K. J. (2010). Dynamic performance metrics to assess sustainability and cost effectiveness of integrated urban water systems. *Resources, Conservation and Recycling, 54*, 719–736.

Finney, K. N., Chen, Q., Sharifi, V. N., Swithenbank, J., Nolan, A., White, S., & Ogden, S. (2012). Developments to an existing city-wide district energy network: Part II—Analysis of environmental and economic impacts. *Energy Conversion and Management, 62*, 176–184.

Fuller, J. (2009). *Heads, you die: Bad decisions, choice architecture, and how to mitigate predictable irrationality*. Surrey Hills, NSW: Per Capita Australia Limited.

Giradet, H. (2004). Metabolism of cities: from creating sustainable cities. In S. M. Wheeler & T. Beatley (Eds.), *The sustainable urban development reader* (pp. 125–132). New York: Routledge.

Glaeser, E. (2011). *The triumph of the city: How our greatest invention makes US richer, smarter, greener, healthier and happier*. Oxford: Macmillan.

Glaeser, E., & Kahn, M. (2008). *The greenness of cities: Carbon dioxide emissions and urban development*. Working Paper, Havard Kennedy School, Taubman Centre for State and Local Government.

Gore, A. (2009). *Our choice: A plan to solve the climate crisis*. New York: Penguin Group.

Griffiths-Sattenspiel, B., & Wilson, W. (2009). *The carbon footprint of water*. Retrieved from http://www.csu.edu/cerc/researchreports/documents/CarbonFootprintofWater-RiverNetwork-2009.pdf.

Hall, M., West, J., Lane, J., de Haas, D., & Sherman, B. (2009). *Energy and greenhouse gas emissions for the SEQ water strategy*. Retrieved from http://www.urbanwateralliance.org.au/publications/UWSRA-tr14.pdf.

Hatt, B. E., Deletic, A., & Fletcher, T. D. (2005). Integrated treatment and recycling of stormwater: A review of Australian practice. *Journal of Environmental Management, 79*, 102–113.

Hughes, R., Anda, M., Ho, G., & Mathew, K. (2006). Integrated waste management. In D. Marinova, D. Annandale, & J. Phillimore (Eds.), *International handbook on environmental technology management* (pp. 405–421). Cheltenham: Egward Elgar.

Intergovernmental Panel on Climate Change (IPCC). (2007). Climate change 2007: Mitigation of climate change. In B. Metz, O. Davidson, P. Bosch, R. Dave, & L. Meyer (Eds.), *Contribution of working group III to the fourth assessment report of the intergovernmental panel on climate change*. Cambridge: Cambridge University Press.

Jackson, T. (2005). *Motivating sustainable consumption: A review of evidence on consumer behaviour and behavioural change*. Surrey, UK: Sustainable Development Research Network.

Jelle, B. P. (2011). Traditional, state-of-the-art and future thermal building insulation materials and solutions—Properties, requirements and possibilities. *Energy and Buildings, 43*, 2549–2563.

Kenway, S. J., Priestly, A., Cook, S., Seo, S., Inman, M., Gregory, A., & Hall, M. (2008). *Energy use in the provision and consumption of urban water in Australia and New Zealand*. Water for a Healthy Country Flagship: CSIRO.

Knights, D., MacGill, I., & Passey, R. (2007). The sustainability of desalination plants in Australia: Is renewable energy the answer? In *OzWater Conference, Sydney, NSW*, March 5, 2007.

Lehmann, S. (2013). Low carbon construction systems using prefabricated engineered solid wood panels for urban infill to significantly reduce greenhouse gas emissions. *Sustainable Cities and Society, 6*, 57–67.

Lou, X. F., & Nair, J. (2009). The impact of landfilling and composting on greenhouse gas emissions—A review. *Bioresource Technology, 100*, 3792–3798.

McIntosh, J., Newman, P., Crane, T., & Mouritz, M. (2011). *Alternative funding mechanisms for public transport in perth: The potential role of value capture discussion paper*. Perth: Curtin University Sustainability Policy (CUSP) Institute.

Medeazza, G., & Moreau, V. (2007). Modelling of water-energy systems: the case of desalination. *Energy, 32*, 1024–1031.

Mees, P. (2010). *Transport for suburbia: Beyond the automobile age*. London, UK: Earthscan.

Moloney, S. E., Horne, R., & Fien, J. (2010). Transitioning to low carbon communities—From behaviour change to systemic change: Lessons from Australia. *Energy Policy, 38*, 7614–7623.

Morrissey, J., Moore, T., & Horne, R. E. (2011). Affordable passive solar design in a temperate climate: An experiment in residential building orientation. *Renewable Energy, 36*, 568–577.

Mountouris, A., Voutsas, E., & Tassios, D. (2006). Solid waste plasma gasification: Equilibrium model development and exergy analysis. *Energy Conversion and Management, 47*, 1723–1737.
Nässén, J., Holmberga, J., Wadeskog, A., & Nyman, M. (2007). Direct and indirect energy use and carbon emissions in the production phase of buildings: An input–output analysis. *Energy, 32*, 1593–1602.
Newman, P. (2006). Transport greenhouse gases and Australian suburbs. *Australian Planner, 43*(2), 6–7.
Newman, P., Beatley, T., & Boyer, H. (2009). *Resilient cities: Responding to peak oil and climate change*. Washington, DC: Island Press.
Newman, P., & Jennings, I. (2008). *Cities as sustainable ecosystems: Principles and practices*. Washington, DC: Island Press.
Newman, P., & Kenworthy, J. (1999). *Sustainability and cities: Overcoming automobile dependence*. Washington, DC: Island Press.
Newman, P., & Kenworthy, J. (2015) . *The end of automobile dependence: How cities are moving beyond car-based Planning*. Washington DC: Island Press.
Newman, P., & Matan, A. (2013). *Green urbanism in Asia*. Singapore: World Scientific.
Newman, P., Rauland, V., Bunning, J., & Beattie, C. (2012). *North Port Quay: A world class carbon neutral development*. Industry Report. Perth: Curtin University Sustainability Policy (CUSP) Institute.
Newton, P., & Meyer, D. (2012). The determinants of urban resource consumption. *Environment and Behavior, 44*(1), 107–135.
Newton, P., Newman, P., Glackin, S., & Trubka, S. (2012). Greening the greyfields: Unlocking the development potential of middle suburbs in Australian cities. *World Academy of Science, Engineering and Technology, 71*, 138–157.
Newton, P. W., & Tucker, S. N. (2010). Hybrid buildings: A pathway to carbon neutral housing. *Architectural Science Review, 53*(1), 95–106. doi:10.3763/asre.2009.0052.
PMSEIC. (2010). *Challenges at energy-water-carbon intersections*. Prime Minister's Science, Engineering and Innovation Council, Canberra, Australia.
Preservation Green Lab. (2011). *The greenest building: Quantifying the environmental value of building reuse*. Retrieved from: http://www.preservationnation.org/information-center/sustainable-communities/green-lab/lca/The_Greenest_Building_lowres.pdf.
Productivity Commission. (2006). *Waste management*. Report number 38. Canberra: Commonwealth of Australia.
Rauland, V., & Newman, P. (2011). Decarbonising Australian cities: *A new model for creating low carbon, resilient cities*. Paper presented at the MODSIM2011: 19th International Congress on Modelling and Simulation, Perth, WA, December 12–16, 2011.
Roseland, M. (2012). *Toward sustainable communities: Solutions for citizens and their governments* (4th ed.). Gabriola Island, BC: New Society Publishers.
Rothhausen, S. G. S. A., & Conway, D. (2011). Greenhouse-gas emissions from energy use in the water sector. *Nature Climate Change, 1*, 210–219.
Seadon, J. (2010). Sustainable waste management systems. *Journal of Cleaner Production, 18*, 1639–1651.
Sharma, A., Burn, S., Gardner, T., & Gregory, A. (2010). Role of decentralised systems in the transition of urban water systems. *Water Science & Technology: Water Supply—WSTWS, 10*(4), 577–583.
Sharma, A. K., Gray, S., Diaper, C., Howe, C., & Liston, P. (2008). Assessing integrated water management options for urban developments—Canberra case study. *Urban Water Journal, 5*(2), 147–159.
Siddiqi, A., & Anadon, L. D. (2011). The water–energy nexus in Middle East and North Africa. *Energy Policy, 39*, 4529–4540.
Stevanović, S. (2013). Optimization of passive solar design strategies: A review. *Renewable and Sustainable Energy Reviews, 25*, 177–196.
Sturgis, S., & Roberts, G. (2010). *Redefining zero: Carbon profiling as a solution to whole life carbon emissions measurement in buildings*. London: RICS Research.

Thormark, C. (2002). A low energy building in a life cycle—Its embodied energy, energy need for operation and recycling potential. *Building and Environment, 37*, 429–435.

Thormark, C. (2006). The effect of material choice on the total energy need and recycling potential of a building. *Building and Environment, 41*, 1019–1026.

Tillman Lyle, J. (2004) Waste as a resource. In S. M. Wheeler & T. Beatley (Eds.), *The sustainable urban development reader* (pp. 133–140). New York: Routledge.

Trubka, R., Newman, P., & Bilsborough, D. (2010a). The costs of urban sprawl-infrastructure and transportation. *Environment Design Guide, GEN, 83*, 1–6.

Trubka, R., Newman, P., & Bilsborough, D. (2010b). The costs of urban sprawl—Physical activity links to healthcare costs and productivity. *Environment Design Guide, GEN, 85*, 1–13.

Trubka, R., Newman, P., & Bilsborough, D. (2010c). The costs of urban sprawl—Predicting transport greenhouse gases from urban form parameters. *Environment Design Guide, GEN, 84*, 1–16.

Wang, Y. D. (2009). Integrated policy and planning for water and energy. *Journal of Contemporary Water Research & Education, 142*, 46–51.

Western Australian Land Information Authority. (2012). *A guide to strata titles in Western Australia*. The Government of Western Australia, Perth. Retrieved from https://www.landgate.wa.gov.au/docvault.nsf/web/PS_STPM/$FILE/STRATA-Title-Guide-ONLINE.pdf.

Wolman, A. (1965). The metabolism of cities. *Scientific American, 213*(3), 179–190.

Zuo, J., Read, B., Pullen, S., & Shi, Q. (2012). Achieving carbon neutrality in commercial building developments—Perceptions of the construction industry. *Habitat International, 36*, 278–286.

Chapter 13
Making It Work

13.1 Introduction

Various forms of low-carbon and carbon-neutral urban development are being demonstrated around the world. As outlined in Chap. 6, these developments provide multiple benefits to their residents and to the wider city, as well as the planet. However, this form of development has not yet become mainstream despite the numerous benefits it offers. The previous chapter showed that it is possible to provide certification for such development, though it will need specific rules, guidelines and governance systems in place. Having a standard way of assessing and certifying this type of development will help to achieve more low carbon and carbon-neutral development. However, there are many additional factors that make it difficult to achieve low-carbon precincts. This chapter attempts to identify and discuss the wider issues that need to be addressed to ensure the decarbonising of cities can be mainstreamed.

The chapter begins by highlighting some of the broad benefits of low-carbon development for both developers and property owners, before examining what is preventing greater adoption of this form of development. Some key obstacles facing developers are identified, such as regulatory and financial barriers. Even without specific barriers, however, the simple 'lack of drivers for change' (Zuo et al. 2012: 279) has been identified as a key reason why developers continue to adopt a business-as-usual (BAU) approach. Roseland (1997) identifies the general lack of guidance in assisting with the practical application of low-carbon development at the community level.

These issues make it clear that this sector currently requires greater assistance and support in order to facilitate increased uptake of low-carbon design and construction within the development industry. Assistance may be in the form of financial incentives, as well as new governance approaches and 'support mechanisms…to provide reliable guidance on how to incorporate sustainability into practice'

V. Rauland and P. Newman, *Decarbonising Cities*, Green Energy and Technology, DOI 10.1007/978-3-319-15506-7_13

(Jones and Patterson 2007: 258). Several examples of these types of opportunities are highlighted, many of which may only be required in the short term, until such developments begin to become more mainstream.

13.2 Benefits of Low-Carbon Development

The benefits for people purchasing properties or renting within low-carbon developments include the following:

- a reduction in energy and water consumption and thus utility costs;
- reduced need for car-based travel and thus reduced fuel costs;
- reduced overall emissions and carbon footprint, and;
- increased liveability from better access to amenities and shorter commuting times.

There are also many advantages for developers pursuing this form of development, which include the following:

- the market advantage of improved product positioning that can help to differentiate them from competitors;
- reduced risk associated with carbon pricing and GHG reporting, as well as from future legislation around energy efficiency and stricter building codes;
- reduced construction costs[1];
- opportunities to promote corporate social responsibility, and;
- the ability to demonstrate a company's commitment to the environment and addressing climate change.

Furthermore, benefits also accrue to the broader economy, as developers who pursue low-carbon design and construction have the ability to influence their entire supply chain, thus creating a shift in thinking towards 'low carbon' within a variety of industries. This shift towards low-carbon products and services will help to drive the new green economy, which is rapidly emerging (Newman and Kenworthy 2015).

13.3 Barriers to Uptake of Low-Carbon Development

The knowledge and technology required to decarbonise our built environment already exists. There are also a variety of policies and initiatives in place to promote the shift to these low-carbon options. However, despite the opportunities and incentives, progress and uptake of low-carbon urban development by developers

[1] For example, if developers use prefabricated materials, which can be cheaper to produce and reduce the construction time, as discussed in Chap. 12.

appears to be relatively slow. The largely unconnected nature of the existing opportunities, policies and initiatives is argued to play a significant role in obstructing progress by creating confusion. Several other obstacles have also been identified in the literature as inhibiting greater adoption of low-carbon and energy efficiency measures within the built environment, including the high capital cost of energy efficiency upgrades, split incentives and regulatory barriers. These and other barriers are discussed below.

13.3.1 Information Barriers

Information barriers affect numerous areas of the economy, distorting and often negatively affecting decisions made, thus leading to various market failures (Garnaut 2008). Information barriers are often discussed in relation to energy efficiency and commonly refer to the situation where a lack of appropriate information prevents consumers from making optimal purchasing choices, thus preventing the market from operating most efficiently. However, the lack of information can affect more than just the final purchasing decision. It can affect decisions being made along the entire supply chain, thus affecting the overall outcome or development of a product or service.

In relation to low-carbon urban development, information barriers exist at all levels of the development process, from the design and construction of developments and the marketing and selling of properties, to the operation of houses, buildings and appliances by owners and the lifestyle options adopted by residents and occupants. These are discussed further below.

13.3.1.1 Developers/Builders

Developers currently lack information about low-carbon infrastructure options, including the carbon and costs associated with different options.[2] This prevents builders, designers, architects and planners from making optimal choices based on quantifiable numbers. It is unrealistic to commit any developer to reduce carbon or go carbon neutral without knowing the costs and carbon implications for a project. Thus without easily accessible information in these two important areas, developers are likely to make arbitrary choices or continue to follow BAU approaches (Newman et al. 2011). The desire to maintain BAU has resulted in a significant lack of innovation within the mainstream construction industry, particularly around sustainability (Glass et al. 2008).

---

[2] The issue of long-term versus short-term costs and benefits are discussed further in Sects. 13.3.2 and 13.3.3.

A study of the perceptions of construction industry professionals has also demonstrated the lack of understanding by those working in the field about what low carbon and carbon neutral actually mean in the context of the built environment, particularly given the lack of clear definitions available (Zuo et al. 2012).

Another information barrier relates to the lack of knowledge and information by tradespeople, engineers and subcontractors about green design and technologies, which hinders their ability to deliver the new low-carbon Green Infrastructure desired, or required, by developers or governments (Glass et al. 2008; Hwang and Ng 2013; International Labour Office 2011; Williams 2012).

13.3.1.2 Marketing Low-Carbon Developments

At the marketing level, adequate information is often lacking about the benefits of green or low-carbon designs, technologies and construction, which leads to serious market failure. For example, if real estate agents or marketing agencies fail to identify or understand the benefits associated with key sustainability features, these features are unlikely to be priced into the properties, despite the additional costs associated with them. Technologies such as PVs and batteries cost more upfront but can be paid off in a few years, after which time households can save considerable money. Such information needs to be a part of real estate marketing.

Developers need to be able to pass on the costs associated with improved environmental and operational performance to consumers in order to remain viable (Ball 2002). Being able to identify, discuss and promote the sustainability features of developments and buildings is likely to require additional training and education within the real estate sector (Pacifici 2011).

13.3.1.3 Consumers/Occupiers

As previously discussed, consumers are often not provided with enough information about the benefits and limitations of various building designs or features, particularly in relation to long-term versus short-term benefits. The result is a reluctance of consumers to pay more for green features.

Another barrier for consumers in making appropriate sustainable purchasing decisions concerns their ability to understand complex information. The public does not always have an adequate background education to understand the specific features of environmental alternatives, especially if they involve complex technical or engineering-related information. In many cases, labelling and certification schemes are intended to fill this gap (e.g. Green Star Certification, energy efficiency ratings on appliances and ISO standards for sustainable construction), though issues still abound regarding the credibility of some schemes (Ball 2002).

13.3.1.4 Circle of Blame

A barrier to the construction of more sustainable buildings and urban development involves what has been termed the 'vicious circle of blame' (Sedlacek and Maier 2012). This negative feedback loop is as follows. Building occupiers say they would purchase more sustainable buildings, but they are not offered to them by the market; builders claim they could easily build more sustainably, but developers do not request them to do so; developers argue that investors do not require them; and investors claim there is no demand for them by the market. The situation therefore is a self-perpetuating cycle of blame.

Sinclair (2008) proposes that several low-carbon demonstration projects are needed in order to break the cycle, noting that 'widespread development of innovative carbon-neutral projects of varying scales should not be expected until a precedent has been set, and a planning framework has been tailored to exploit the benefits of this in future development projects' (61).

Having sufficient market demand for carbon-neutral development projects was also the number one 'critical success factor' identified by Zuo et al. (2012) in their research into carbon neutrality in commercial building developments (see Fig. 13.1).

13.3.1.5 Addressing Information Barriers

Governments are beginning to address some of the issues associated with information barriers. The Australian government's mandatory energy disclosure for

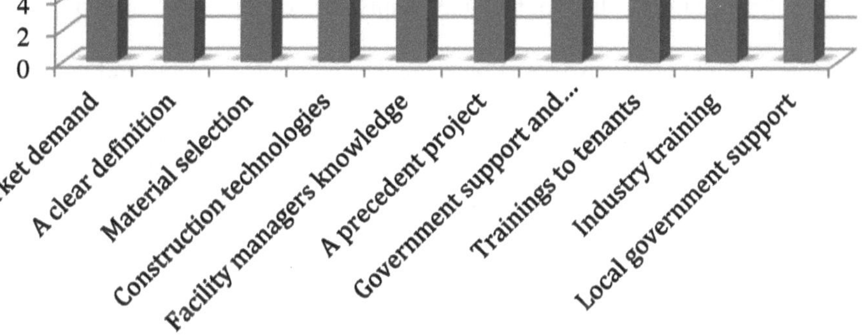

Fig. 13.1 Critical success factors for carbon-neutral commercial building developments. *Source* Adapted from Zuo et al. (2012)

commercial buildings has been a positive step forward in addressing a lack of information relating to energy efficiency in buildings. This sort of information can help to promote more energy-efficient, low-carbon buildings and is discussed further in Sect. 13.5.

Other options for overcoming information barriers that are being actively pursued by government relate to behaviour change. Even when a consumer has bought an environmentally friendly product (be it an appliance or a home), they will often need appropriate advice and information on how best to operate it. These sorts of issues are being addressed through various behaviour change programmes worldwide such as Living Smart and Travel Smart, discussed in Chap. 12. An overview of behaviour change programs in this area is provided by Fudge and Peters (2014).

Overcoming information barriers at all the levels discussed above is likely to require a range of new skills and knowledge by all participants, as the absence of these skills and knowledge significantly inhibits a low-carbon agenda (DCCEE 2010). Greater technical expertise and specialist knowledge in areas related to improved energy, water and waste efficiency are particularly important within the construction industry. The creation of this new knowledge and skills will require training, thereby creating a variety of new green jobs as a vital component of the new green economy.

13.3.2 High Capital Costs

Low-carbon developments generally have had higher upfront costs associated with them,[3] as they entail new and emerging technologies and greener materials that are more expensive (International Labour Office 2011). Although the higher upfront costs are usually balanced out by long-term savings, the initial investment is still perceived to be risky and a barrier by the development industry, which worries that the costs associated with the long-term benefits may not be able to be passed on to the building owner. The problem posed by short-term costs and long-term benefits is further discussed below under split incentives. New technologies also carry greater financial risk due to the uncertainty associated with their performance (DCCEE 2010).

However, the costs associated with greener building products and technologies do decrease in time as increased demand creates economies of scale and brings new producers into the market—thus driving competition (Wills 2011).[4] Nevertheless, it is argued that, initially, new funding mechanisms and incentives are needed to assist

---

[3] It is acknowledged that this is not always the case. For example, studies of prefabricated wooden panels used in construction of low-carbon buildings have usually shown reduced costs compared to BAU, but not easily adopted (Lehmann 2013).

[4] See also http://www.futuresmart.com.au/rtwtechadopt.html.

the construction and development industry in pursuing these low-carbon options. Some suggestions for this are proposed in Sect. 13.4.

13.3.3 Split Incentives

The notion of split incentives was discussed briefly in Chap. 6. It involves a situation where the costs of an action (e.g. an energy-efficient upgrade) are borne by one party (such as the building owner), while the benefits accrue to another party (for example, the tenant) (WBCSD 2009). In terms of low-carbon urban development, the critical issue faced is that the generally higher upfront costs associated with Green Infrastructure are borne initially by the developer, while the residents receive the long-term benefit. This in itself is not a problem, if a developer can pass on the higher upfront costs associated with the Green Infrastructure. However, if the owner occupant is not provided with enough information to convince them of the financial and other benefits of Green Infrastructure, which is often the case, they will be reluctant to pay the higher upfront costs needed to cover the developer's additional costs. This creates a lot of risk for developers trying to sell higher quality, low-carbon developments, thus creating an incentive to simply provide cheaper, less efficient developments that are more likely to be sold.

Another important issue is the *ability* of property buyers to afford the higher upfront costs. Purchasers may be well-aware of the long-term benefits but not have enough capital to meet the additional upfront costs associated with more efficient products (in this case a house) and are, therefore, forced to accept a more inefficient alternative. Banks are often very conservative in their lending processes and thus do not easily lend money to green risky developments unless the applicants are wealthy. This has the potential to lead to the development of eco-enclaves, whereby only rich people can afford to live in more sustainable (and ultimately more affordable) developments. An opportunity for dealing with this is provided in Sect. 13.5.6.

Providing access to information about affordable housing within low-carbon urban developments is therefore a critical issue to address and an essential element in the provision of long-term sustainability in our cities.

13.3.4 Longer Approvals Process

Developments that demonstrate innovative low-carbon technologies, new efficient Green Infrastructure and alternative designs are more likely to require longer periods for development approval (PIA 2011), which ultimately represents additional costs to a developer. The financial risk borne by developers and banks on such green developments can be eased by planning agencies having access to the same tools as developers that can quickly show carbon and cost implications in their planning options and hence hasten decisions.

13.3.5 First-Mover Disadvantage

Also referred to as the free-rider problem, the first-mover disadvantage occurs when a person or business shoulders large initial costs associated with introducing a new technology or concept that subsequent businesses may benefit from (DCCEE 2010; Garnaut 2008). In terms of the development industry, an example could be a developer having to jump through a variety of regulatory hoops in order to get a new technology approved, and thus facing significant additional costs associated with the delayed approvals process. Subsequent developers, however, may benefit from the precedent set by the initial developer, thus shortening approval times and cutting costs. This is not a barrier to mainstreaming low-carbon development, as long as the information from the first-mover advantage is freely available and accepted by the next movers. Some options to deal with this include the removal of stamp duty, land tax exemptions and a fast track approvals process, which are discussed in more depth in Sect. 13.5.4.

13.3.6 Policy and Pricing Uncertainty

Every country finds carbon pricing difficult and the politics is often very combative, creating uncertainty in the market, including in the land development market. The introduction of a price on carbon in Australia in July 2012 was designed to bring certainty to business and industry and to encourage investment in low-carbon options.

However, an election in 2013 saw a new Coalition Government replace the previous Labour Government and reintroduced political uncertainty around carbon policy. The new government repealed the carbon pricing legislative package, demonstrating that carbon pricing in Australia remains politically divided. The repeal came despite a report (see Clean Energy Future 2013)[5] being released on the carbon price (after one year of operation), which showed it had been successful in reducing emissions and that the economy had not been damaged in any way. The report showed that the country had increased power generation from renewable sources by 28.5 %, is more energy efficient, and has reduced its carbon emissions. There was a drop in electricity generated from black coal and brown coal, 4.2 % and 13.3 % respectively, from the previous period before the tax (2011–2012) (DIICCSRTE 2013).

There is some question as to whether the loss of a carbon price signal will mean a return to more profligate energy usage and reduction in the growth of renewables in Australia. Internationally, the signals to move away from the fossil fuels are now finally entrenched in the global economy and most businesses and households are aware of the advantages of the new global green economy. Fortunately, renewables

---

[5] The report titled "How Australia's carbon price is working—one year on," was released by Australia's Department of Industry, Innovation, Climate Change, Science, Research and Tertiary Education and used in Australia's communication to the UNFCCC on how Australia is meeting its target.

and energy efficiency pricing are being driven more by China than by a local carbon tax mechanism (Sheehan et al. 2014).

In terms of urban development, however, the Abbott Government's Direct Action Plan opens up the possibility of generating carbon credits from low-carbon precincts and urban development, which as we have discussed, was previously ineligible. It thus offers an example for an End-User approach to carbon credits that could help define a new approach to mainstreaming carbon reductions in the built environment.

Globally, there is less political uncertainty, as most countries are moving towards carbon pricing.

13.3.7 Lock-In

Infrastructure lock-in is a serious issue and a significant barrier for retrofitting low-carbon infrastructure in existing cities. Large infrastructure generally requires large capital investments and thus has relatively long lifetimes (often 50–100 years). Examples of infrastructure with long lifetimes include large-scale coal-fired power plants, freeways and in more recent times, desalination plants. Decisions to build low-density housing developments on the fringe of cities are also locking-in patterns of infrastructure that will be hard to retrofit into the future (Newton 2008). Indeed, all these infrastructure choices have shaped Australian cities for more than half a century (ARUP and C40 Cities Climate Leadership Group 2011), and most have considerable emissions attributed to them. They are thus locking substantial emissions into cities for a significant time period. These decisions will have a significant effect on the ability to deliver low-carbon urban development within our cities and highlight the need to ensure that future low-carbon built form is delivered in a way that is adaptable and easily replaceable when newer, cleaner technologies and urban systems are developed.

Another issue to consider is the vulnerability to the impacts of climate change that is essentially being locked-in from various infrastructural systems. Kennedy and Corfee-Morlot (2013) suggest that 'it may be a cost-effective decision not to "overinvest" to boost resilience in an infrastructure system' (2). Small-scale distributed systems, which require lower capital investment, will be easier to retrofit in the future when newer, more efficient technologies become available.

13.3.8 Credibility of Carbon Claims

As noted throughout this book, credibility of low-carbon and carbon-neutral claims for development remains a key issue in relation to wider adoption. Even those working within the industry are dubious of the concept, with key stakeholders identifying 'an initial lack of accreditation of the concept and workable measures for carbon reduction' as an impeding factor preventing greater adoption (Zuo et al. 2012: 282).

13.3.9 Multiple Stakeholders

As highlighted in Chaps. 9 and 11, the design and construction of precinct-scale urban development has numerous stakeholders that, if not managed well, can act as a barrier to achieving low-carbon or carbon-neutral land development. Success requires significant collaboration between the various stakeholders. Blundell (2012) identifies utilities as an important stakeholder, arguing 'there also should be more coordination between those in the property industry who are planning and developing more resilient communities, and the utilities sector' and noting that without their support, low-carbon developments will fail.

13.3.10 Regulatory Issues for Utilities

Utility regulatory barriers pose a significant problem for low-carbon urban development. As highlighted in Chap. 4, new low-carbon infrastructure is likely to increasingly be based around decentralised or distributed systems for resource management. However, given that cities are currently dominated by centralised infrastructural systems, most of the governing regulations are designed for this level of supply. Consequently, various regulatory issues currently prevent or make it difficult for small-scale energy generators to enter the market and access the electricity grid. There are also technical issues that currently prevent distributed energy being fed into the grid, though such issues could be overcome with the introduction of a smart grid. Bunning et al. (2013) discuss some governance issues currently affecting decentralised, low-carbon infrastructure and identify emerging governance models that deal with resources at a more local level.

Mobbs (2010) highlights a range of regulatory issues, which, although they may not necessarily block the installation or implementation of new low-carbon infrastructure, can act as a disincentive preventing developers from pursuing low-carbon development. An example he cites is the requirement for developers to pay for centralised services (e.g. connection to sewerage systems) even if they do not connect to the systems (Mobbs 2010).

There are various other water-related regulations, particularly around health, that inhibit the uptake of small-scale, onsite wastewater recycling. Toze (2006) identifies that overly strict health regulations can result in 'the requirement for over-treatment of recycled water, which can lead to potentially valuable water reuse projects becoming too costly and inefficient to be viable' (41).

Dealing with these obstacles can be difficult and is likely to require new governance and participatory models specially designed for the development industry (Bunning et al. 2013). These structural changes have been associated with each wave of technological and economic innovation, as shown by Newman et al. (2009) and should not be held back as they will enable the next wave of productivity and liveability to be adopted.

13.4 Opportunities to Overcome Barriers

Existing examples of low-carbon precinct-scale development around the world have generally required significant support in order to get off the ground and thus are generally seen as 'demonstration projects' (see Thomson et al. 2013), still situated in the early innovation stage within the innovation framework as illustrated in Fig. 13.2.

A study examining business models for enabling sustainable precincts in Australia found that few low-carbon developments were produced in a commercially viable way (Moreland Energy Foundation et al. 2011). It revealed that most developments are still far from being mainstream and require various forms of assistance, support and/or incentives in order to make them viable. Thomson et al. (2013) also found this in their review of low carbon precincts. This is due to the various barriers identified above. In European nations such as Germany and Denmark, these barriers are being dismantled much faster and hence green economy urban innovations are far more common (Lehmann 2015). The arrival of China into this market is hastening the changes even faster (Newman and Kenworthy 2015; Lehmann 2015; Sheehan et al. 2014).

Ensuring a successful transition to low-carbon infrastructure within precincts and cities will no doubt require a range of incentives, changes in regulations and information campaigns. All cities are now competing on this agenda. The section below identifies some opportunities to support the development of low-carbon and carbon-neutral developments, from planning and regulatory assistance to various financial incentives.

13.4.1 Government Leadership and Facilitation

As outlined above, global leadership in low-carbon, eco-precincts appears to be happening in Europe, with China quickly following along this path. In places such as

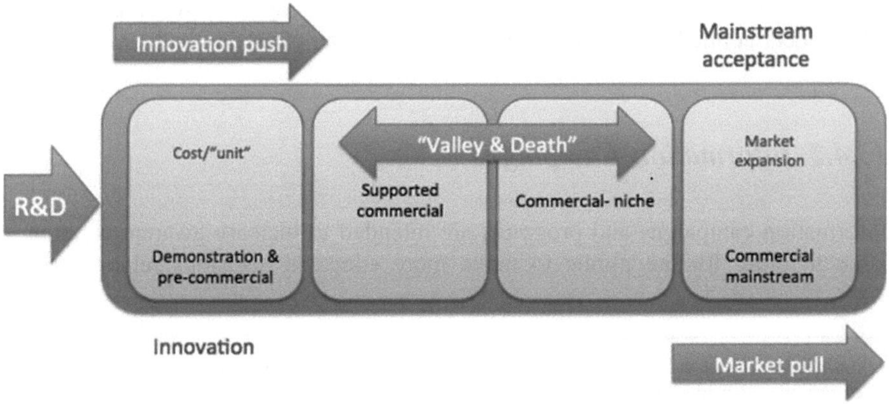

Fig. 13.2 Innovation framework. *Source* Adapted from Moreland Energy Foundation et al. (2011)

Australia and America, innovations around low carbon in urban development have mostly come from local government (Newman and Kenworthy 2015). Government leadership and assistance is required to facilitate and enable low-carbon and carbon-neutral developments, particularly while they are still in the niche stage of the innovation cycle (Moreland Energy Foundation et al. 2011; Bunning et al. 2013).

The city of Sydney in Australia provides a good example of government leadership. In their drive to reduce local emissions by 70 % by 2030, they introduced a Tri-generation Master Plan, which they are now working towards implementing in multiple locations across the CBD. These systems will provide local precincts with low-carbon and low-cost energy, and carbon-free heating and cooling. As many barriers stood in the way of implementing this small-scale technology, the city of Sydney has had to pioneer this space, negotiating with local energy companies and facilitating the changing of rules and regulations around energy supply networks and energy-related services (i.e. heating and cooling) in Australian cities.[6] While they are still working towards this and still having to overcome regulatory hurdles, this will help to enable other demonstrations to occur more easily. This work was based largely on the pioneering work undertaken by the town of Woking in the UK.

Places Victoria, a government land development agency in Victoria, Australia, is another example of a government agency providing leadership. Places Victoria was the first state government agency to successfully implement tri-generation across multiple land titles, thus enabling Australia's first 'precinct-scale' tri-generation system to occur (Revitalising Central Dandenong n.d.).

Other ways government can provide leadership are through adjusting planning regulations and legislation. An example of this was the City of Fremantle in Western Australia removing the requirements for developers in the city's CBD to provide car parking in new developments (Zaw 2012).[7] This enables developers who wish to reduce the development's transport-related emissions to do so. The council has also passed a policy that will set aside dedicated bays on the street for car-share vehicles to incentivise private car-share companies to set up business in the city. Having a car-share program located close to new urban developments will reduce the perceived risk by developers of providing car parking-free developments. Over 1100 cities around the world now have car-share schemes, many of which local councils are either facilitating or leading (Shaheen and Cohen 2013).

13.4.2 Information Campaigns

Information campaigns and programs are intended to increase awareness about an issue in order for the public to make more educated decisions relating to that

[6] Unlike other cities around the world, particularly Scandinavian ones, Australia has no precedent or experience in providing district heating as a service.

[7] Previous requirements were 1–2 car parks per dwelling.

issue.[8] Many successful behaviour change information campaigns and programmes have been developed and run around the sustainability issues in recent years. For an overview of these, see Fudge and Peters (2014), who sets out programs in Europe and Australia similar to the Travel Smart and Living Smart programs discussed in Chap. 12. Such programs have experienced broad success largely due to their tailored and interactive approach. Research has long questioned the effectiveness of broad, untailored media campaigns (McKenzie-Mohr 2007; Cones and Hayes 1980).

Information campaigns are often led by government, but can also be driven by community,[9] businesses or industry, all of which may have some form of vested interest or benefit in some way. Indeed, 'information campaigns can be used to generate markets' (Williams 2012: 164), which benefit businesses entering that market. While businesses can provide valuable marketing campaigns that raise consumers awareness about particular issues, it is also important to have unbiased source of information delivered to consumers and the public.

Garnaut (2008: 409) identifies three key points that should be considered when designing an information campaign. Such a campaign should:

- attempt to overcome biases by providing a simple comparison between current and future costs and current and future benefits;
- use familiar language (such as payback periods); and
- be located as close to the point of sale as possible.

A report of the Prime Minister's Task Group on Energy Efficiency (DCCEE 2010) provides several other key ideas, recommendations and examples of information campaigns. Providing this sort of information to consumers or potential buyers of low-carbon developments could significantly alter how people view properties.

13.4.3 Knowledge Sharing—Both Success and Failure

The lack of knowledge transfer around what has and has not worked within innovative, low-carbon urban development projects can prevent greater learning among the development and property industry and thus reducing the ability to build on past experiences. While sharing success is vital and can help promote new technologies and designs, sharing the lessons learnt from what has not worked and why is just as important to ensure that mistakes are not replicated (Rauland and Thomson 2015).

---

[8] Smoking is an example of an information campaign that has been very successful in demonstrating the effects smoking can have on your health.

[9] Living Smart was initially developed through collaboration between a local community, a university and council. The community continues to play a large role in the ongoing running of the program.

Jones and Patterson (2007) note that there is a tendency among green building projects to emphasise successes and hide failure to avoid backlash, thus 'green-washing' projects. However, this can often lead to a sense of cynicism around sustainability if rumours of failure later emerge.

The Carbon Disclosure Project, which launched a city-based disclosure project in 2008, demonstrates the value in disclosing information particularly regarding strategies cities are putting in place to address emissions and sustainability (Accenture n.d.). This public platform has provided a great opportunity for cities around the world to share knowledge on the performance and success of various strategies and measures, which has not only provided a way for them to communicate their value to stakeholders, but also offers other cities the opportunity to learn from them.

Providing a platform whereby developers, Energy Services Companies (ESCos), local governments and other stakeholders can openly share knowledge and experience around new technologies, planning guidelines and various precinct-scale innovations, would go a long way in progressing the transition towards low-carbon cities.[10]

13.4.4 Creating Baselines for Sector

Developers currently lack meaningful baselines by which to compare their developments at the precinct level. While several baselines and benchmarks exist for buildings, transport, waste and water, nobody really knows what the combined effect of these various sectoral emissions are at the precinct scale, despite the potential for developers to influence the outcomes of each element.

Thus, creating baselines for this sector would be very useful, not least because it would show the difference between current and best practice (in terms of GHG emissions). Furthermore, having well-established baselines can enable a variety of incentives and disincentives to be applied to developers. For example, a baseline and credit scheme specifically designed for urban development could encourage developers to be below (that is, better than) the baseline by allowing them to sell carbon credits as a result, while developers above the baseline must purchase credits to meet their liability. Creating different types of baselines for different urban typologies based on a mixture of features would provide valuable information for comparisons and general knowledge sharing.

[10] It is worth acknowledging that developers may also be reluctant to share some learnings or strategies, if these have given them a commercial edge or advantage. Strategies should be sought to overcome this issue.

13.4.5 Energy Market Reform

As mentioned in Chap. 4, the majority of cities worldwide are designed around large-scale centralised energy generation and distribution networks. Energy regulations and markets have, therefore, been designed to meet and best serve the operational requirements of this large-scale generation. As a result, the current regulatory framework has created many barriers preventing small-scale, distributed energy generators from easily entering the market (Garnaut 2008). A report by CSIRO (2009), which analysed the value of distributed energy in Australia, identified numerous policy and regulatory barriers (among other types of barriers)[11] that are currently inhibiting uptake of distributed energy in Australia. Considering the substantial investment in the infrastructure associated with current energy networks, it is not surprising that there is reluctance, particularly by those with vested interests, to make 'broad regime change that threatens viability of incumbent assets' (Szatow et al. 2012: 2). There is considerable evidence emerging of stranded assets in centralised coal-fired generation systems.[12]

Energy market reforms are critical and will be essential in the long term to facilitate and encourage greater adoption of decentralised energy into the national energy market (Garnaut 2008). Szatow et al. (2012) also suggest, however, that it may be worthwhile investigating and pursuing alternative ways of addressing energy, such as through heating, cooling and energy efficiency services as opposed to electricity provision, by building those services into, and providing them as part of, the property development industry. These reforms are likely to emerge when precinct-scale innovation and new tools for assessment become more mainstream.

13.4.6 Alternative Funding Models to Assist Low-Carbon Urban Development

Various funding models and solutions have emerged in recent times to deal with some of the barriers discussed in the previous section, particularly high upfront capital costs and split incentives (Rauland 2014). Although some were initially designed to assist sustainable investment in single ownership buildings, several options could potentially be expanded to enable the delivery of district-scale eco-infrastructure. To do so effectively, however, is likely to require new governance processes. Box 2 puts forward a potential new governance mechanism that could help to manage such financing arrangements and facilitate the delivery of eco-infrastructure (GRIDs).

---

[11] These include financial costs (e.g., upfront equipment costs, access to market and energy prices), decision-making, energy market structure (e.g., utilities, service provides, distribution companies), information and technical barriers.

[12] See http://www.abc.net.au/radionational/programs/scienceshow/24-hr-solar-power-on-the-horizon/5980972.

Some of the financing options include the following: Energy Service Companies (ESCOs), which potentially shifts the capital cost burden away from developers or local councils to a private company who installs and manages the infrastructure; Green Bonds or Climate Bonds to fund Green Infrastructure projects; Property Assessed Clean Energy (PACE) financing (primarily in the US) or Environmental Upgrade Agreements (EUA) (in Australia) to assist with upgrades and retrofits in existing buildings; community-owned renewable energy projects; and leasing arrangements, particularly for solar panels. These and others funding models are examined in Thomson and Rauland (2014). EUAs and ESCos are discussed further below.

13.4.7 Environmental Upgrade Agreements

EUAs are an Australian adaptation of the PACE financing mechanism originally developed in the USA (DCCEE 2010). While the PACE mechanism initially targeted homeowners, EUAs focus on commercial buildings (Blundell 2012). EUAs allow the cost of an energy-efficient retrofit or upgrade of a building to be attached to the property rather than being borne solely by the building owner, thus allowing the cost to be passed on the next owner. The reason why building owners have been reluctant to pursue retrofits in the past was the concern that they would not be able to pass through the additional costs associated with the upgrade to the next buyer if/once they sold the building, or indeed, to the tenant who actually receives the benefit from the lower operation energy costs.

The EUA mechanism allows low-cost loans to be taken out by the building owner and used to retrofit a building, with loans being repaid through an agreed special council rate attached to the building over a set period of time. This allows the cost associated with the retrofit of building to be passed on to future building owners (Blundell 2012). This pioneering initiative has helped to overcome the split incentives problem facing existing commercial buildings.

While this issue is perhaps less relevant to developers who are usually building new developments (or redeveloping existing sites), as opposed to upgrading existing buildings, the idea of attaching the costs to a property rather than an individual owner/developer has merit and is worthy of further investigation. For example, if a developer agrees to install co- and tri-generation infrastructure within a development, though is not responsible for the implementation or operation of the actual system, this additional expenditure (which is far more cost-effective to install during the development stage rather than retrofitting pipes at a later stage) could be a cost that is attached to the property and repaid through a fixed term, council loan, over a longer period (i.e. 15 years). This idea is further discussed in the section below.

13.4.8 ESCo's

An Energy Service Company, or ESCo as they are commonly referred to, is another mechanism that can help to overcome barriers associated with implementing energy efficiency upgrades and decentralised renewable or low-carbon energy generation projects. ESCos differ from standard utility companies by offering energy services (e.g. thermal comfort or improved lighting) as opposed to solely providing electricity. ESCos are particularly useful in dealing with the high upfront capital costs associated with energy-related projects and removes the risk associated with the uncertain energy payback periods (Szatow et al. 2012).

An ESCos business model generally involves the ESCo paying the initial purchase and installation costs associated with energy-efficient equipment, appliances and/or systems, with this investment/loan being repaid through the energy savings resulting from the upgrades within the building over a set period of time. The repayment amount is determined by the ESCo, and while it usually equates to slightly more than the initial appliance purchase cost, the overall benefit it brings to the building owner by removing the upfront capital investment cost usually outweighs this additional cost spread over time. After the payback period, the building owner benefits from reduced ongoing utility bills.

The scope of services that ESCos provide varies from lighting retrofits to installing and managing small- to medium-scale onsite energy generation, though ESCos are often contracted to identify a range of energy efficiency solutions in a building (ASBEC 2008). The ESCo usually manages and maintains the equipment during the entire payback period to ensure that the equipment is operating as efficiently as possible in order to maximise the energy savings.

In terms of precinct-scale developments, the ESCo model is particularly appealing as it functions predominantly at the small, decentralised level discussed throughout this book. At the precinct level, an ESCo could be responsible for building and operating a co- or trigeneration system. It could also be used in combination with a variety of measures identified in this section, such as the GRID mechanism mentioned in Box 2.

13.4.9 Business Improvement Districts for Precinct Development

The Business Improvement District (BID) concept emerged first in Toronto, Canada, in the 1970s and has grown considerably since with estimates of up to 60,000 BID-like schemes now existing worldwide (Hoyt 2004). A BID is essentially a body created and funded by local businesses and property owners within a defined district of a city, and it is responsible for managing the improvement of that district. A BID is most often funded by an increase in tax or a levy applied to businesses and property owners situated within the area of the BID. The stakeholders will have identified a need for revitalisation within their area and will generally be involved in the formation of it and therefore are willing to pay the extra money in order to fund the process. Some BIDs also accept donations (Levy 2001).

BIDs are often responsible for managing (but not necessarily delivering) services such as street cleaning, safety and security, as well as coordinated local marketing campaigns for businesses, in order to increase the attractiveness, and thus patronage, of local downtown business areas. Nevertheless, the roles and functions of BIDs vary considerably between different districts, cities and countries, depending on the needs of the local area (Levy 2001). BIDs have also been adapted to fit various other districts within cities, such as suburban mixed-use centres,[13] and residential and industrial areas.

BIDs use a very participatory process, whereby the people contributing to the fund play a part in deciding what the funds should be used for. BIDs are thus an alternative, privately funded, independent and participatory governance mechanism for improving specific precincts of cities. The possibility of using the concept of a BID to facilitate the implementation of Green Infrastructure is illustrated in Box 1.

Box 2. Overview of the Greening, Regenerative and Improvement Districts (GRIDs) concept. The concept of a Business Improvement District (BID) can be adapted to provide an ideal mechanism for assisting urban development to adopt the new Green Infrastructure discussed in this book.

Regeneration and improvement of urban areas can be greatly assisted by the implementation of new Green Infrastructure, which will not only help to improve the attractiveness of the area, but also help districts to become more resilient in the face of climate change impacts and rising energy and water prices.

A new term is given to this concept or mechanism—Greening, Regenerative and Improvement District, or GRID, which aptly acknowledges that the new, emerging infrastructure is well suited at a smaller scale—a precinct-scale grid —rather than the traditional centralised grids for managing resources (Rauland 2013). It also acknowledges that to become sustainable, urban areas need to begin producing many more of the resources that they consume. This includes energy, water and potentially food and other resources. This notion of the regenerative or productive city is becoming more widespread.

The issue of engaging multiple stakeholders, which has traditionally posed a problem for urban precinct development projects, is also well addressed through this GRID mechanism, as stakeholder engagement and participation lies at the heart of the BID process.[14]

---

[13] Brisbane City Council introduced a BID-like programme (Hoyt 2004) in Australia in 1996, called Suburban Centre Improvement Projects (SCIPs). This program funded various infrastructure capital works within certain areas, which were paid off using a levy introduced to those areas over a 10-year period (Brisbane City Council 2013).

[14] A new not-for-profit organisation out of the USA, EcoDistricts, has also been calling for the establishment of sustainability management associations within eco-districts or precincts, which would be responsible for the ongoing management of the areas (see http://ecodistricts.org/).

Combining the use of ESCos and various other funding arrangements highlighted above, delivered through a GRID model, could help to provide the basic green urban infrastructure needed at the local precinct level such as co- and tri-generation, community solar PVs and biophilic urban design (including local urban agriculture and precinct-scale storm water management).

This mechanism could potentially help to shift the high upfront capital costs associated with the Green Infrastructure away from the developer, to the community/end-users who will ultimately benefit from the measures. The costs would be spread like a loan over a longer time period and paid through an increased tax or levy within the area. The increased levy would be offset by the reduction in operating costs associated with living and working in those precincts, thus increasing the overall affordability for those buying into the area. As the loan is essentially passed on to future owners/residents (through the annual levy), it also means that the costs are not disadvantaging the initial buyers.

Ensuring an active GRID management association, made up of key stakeholders within the precinct, will be critical to the success of the GRID. Depending on the funds available, this will take the form of an association or a paid organisation (similar to a strata company) and will be responsible for managing the Green Infrastructure and sustainability projects within the precinct. It will also be responsible for seeking out new opportunities for revitalisation and regeneration and as well as new funding opportunities (i.e. grants).

General revitalisation, regeneration and improvement of areas, including the provision of additional activities and amenities can also increase the attractiveness and desirability for living in these areas, which may also help to reduce personal transport-related emissions as people can stay closer to home.

13.5 Mechanisms for Incentivising Precinct-Scale Carbon Reduction

Currently, few incentives exist to reward innovative, low-carbon developments in Australia. Indeed, on the contrary and as demonstrated in this chapter, various obstacles inhibit such developments. The following section provides some examples of incentives that could be offered to low-carbon developments, once they have credibly demonstrated (i.e. through certification) their carbon reductions.

13.5.1 Removal of Stamp Duty

The possibility of removing stamp duty[15] for developments that can demonstrate certifiable carbon reductions could help to incentivise prospective properties owners to buy within low-carbon/carbon-neutral developments. Since such developments may be more expensive in the short term due to the adoption of new technologies, materials, innovative designs and lengthy approval processes, the reduction in cost due to the removal of this fee could help to offset the higher prices, therefore making purchasing into these developments more affordable and attractive. A study by Sayce et al. (2007) demonstrated that stamp duty exemption was one of the options most favoured by industry stakeholders interviewed in the study.

13.5.2 Land Tax Exemptions

Taxes are an economic or market-based instrument designed to address an externality and rectify a market failure (Garnaut 2008). The tax system could be used to incentivise low-carbon development in various ways. Land tax exemptions are just one example of this. Land tax is an annual fee generally paid to state governments by owners of property (Government of Western Australia 2012). The amount required to be paid varies between states and various exemptions usually apply, e.g., for property owners who reside on the property.

Land tax exemptions could be applied to precinct-scale developments that demonstrate credible and quantifiable reductions in GHGs. Various tax concessions, exemptions and incentives are presently being applied to green buildings (DCCEE 2010; GBNYC 2012).

13.5.3 Labelling and Certification Schemes

The concept and use of eco-labelling and certification has grown in line with consumers' willingness to pay extra for products that demonstrate environmental benefits (OECD 2002). Labelling and certification provide a vehicle by which producers (or in the case of this book, developers) can share important information with the public about the environmental attributes and eco-qualities of their product, as these features and processes are often unseen (Brecard et al. 2009). Providing branding opportunities for innovative developers to promote their green credentials is important in helping developers differentiate themselves from their 'business-as-usual' competitors (Bratt et al. 2011). Branding is usually strengthened considerably if it is

[15] Stamp duty is a fee levied on properties at the time of purchase or lease, which is paid to the State.

underpinned by a credible eco-label or certification that demonstrates that the product or development has undergone some form of independent assessment.

13.5.4 Fast-Track Approval Process

Fast-tracking the approval process[16] for 'green' or low-carbon development applications is another form of incentive that can help reward and encourage this type[17] of development. Traditionally, these sorts of development applications have been held up during the approvals stage, which can add significant costs to a development's project budget (Partners in Project Green n.d.). Therefore, creating a process that can expedite the process is a valuable incentive for developers.

The approvals agency can require specific criteria to be met for an application to be deemed 'green' and thus eligible for the fast-track approval process. Alternatively, rating tools can be used to demonstrate that a specific standard has been met. The certification scheme proposed in this book is example of a mechanism that could be used to determine eligibility for this fast-tracking process.

This fast-tracking mechanism has been employed successfully in Vancouver, Canada (Toderian 2013) and in Queensland, Australia (Queensland Government 2011).

13.5.5 Energy Efficiency Standards and Mandatory Disclosure

Mandatory disclosure of energy use in buildings can help bring attention to, and promote, energy-efficient buildings. Germany has had such a policy in place since 2009, whereby all buildings owners are required to provide an energy certificate when selling or leasing property. A similar policy exists in Australia, though currently only for commercial buildings over 2000 m^2 (O'Leary 2012).

The simple act of providing information can help to reward developments that build more sustainable buildings by making their buildings more attractive to purchasers. Without adequate information about the environmental or cost saving benefits of certain features, particularly those related to energy efficiency, higher quality products will continue to be driven out of the market because of a lack of understanding of their benefits (O'Leary 2012).

Precinct-scale mandatory energy disclosure in the design stage, as well as in the sale of the whole precinct, can only happen when modelling tools (such as

[16] The term 'Green Door' has been applied to this concept in Australia.

[17] That is, developments that demonstrate key sustainability, low-carbon or green features.

PRECINX) are fully developed as transparent, and publicly available tools and are required by jurisdictions responsible for planning (Pettit et al. 2014).

13.5.6 Green Mortgages

The concept of a 'Green Mortgage'[18] has been around for several years and has been particularly popular in the United States. A Green Mortgage essentially allows a buyer to take out a larger loan for a house than they could normally afford if the house meets higher energy efficiency standards or has a range of sustainability features that will result in reduced operational costs. Under this scheme, the money expected to be saved on utility bills from a more energy-efficient home is essentially added to a buyer's income and thus valued into the mortgage (Williams 2012).

These types of mortgages can be used for the purchase of both newly constructed energy-efficient houses and older houses that require retrofits. For new construction, the Green Mortgage enables buyers to purchase more expensive properties by taking into account the ongoing reduction in bills. For older houses, the cost associated with energy efficiency retrofits, such as insulation, can be added to the total cost of the mortgage, which again, can be higher based on the expected savings (Deneen n.d.; Munoz 2007).

Green Mortgages could help ease concerns some developers may have that they might not be able to pass on the increased cost associated with greener developments to buyers, as buyers may not be able to afford the higher prices. Green Mortgages could potentially be used as a marketing tool by developers, to increase a buyer's confidence in purchases and increase overall awareness around the long-term benefits of low-carbon design and construction. Banks that provide Green Mortgages can also have a financing equivalent for developers who develop low-carbon precincts, as their resale resilience over time will be higher than developments that are carbon profligates.

13.5.7 Renewable Energy Certificates

Renewable Energy Certificate schemes or Tradable Green Certificates (TGCs) are a policy instrument used to increase the share of renewable energy generated within individual countries and often used as the mechanism to reach renewable energy targets. Various forms of this policy system exist around the world including in Japan, several EU countries (the UK, Italy, the Netherlands and Denmark), over 20 states in the USA, India and Australia (Currie 2013).

[18] Also known as 'Energy Efficient Mortgage' (EEM) or an 'Energy Improvement Mortgage' (EIM).

The Australian Renewable Energy Certificates (RECs) scheme is the mechanism by which the Federal Government expects to meet its Renewable Energy Target (RET) of 20 % by 2020 (Climate Change Authority 2012). This policy was reaffirmed in 2014. The Australian scheme allows two forms of RECs to be produced; those that are generated through the Large-scale Renewable Energy Target (LRET), and those generated through the Small-scale Renewable Energy Scheme (SRES). Developments can potentially generate RECs (most likely SRES) from the renewable energy they produce onsite, which can be traded for cash, but not carbon credits. It is unlikely that an urban development will be able to claim both RECs and funding for carbon abatement under the ERF for renewable energy.

13.5.8 Energy Certificate Schemes

Energy Efficiency Schemes, also known as White Certificate Schemes, are a market-mechanism that currently targets energy efficiency in the residential and commercial building sectors and could potentially target the development industry. Four state-based schemes are currently in operation in Australia (an overview of these schemes is provided in Appendix E) (RedEnergy 2014). There are also several schemes operating internationally. The four Australian schemes are similar in design and generally require energy retailers to produce or purchase, and then surrender, sufficient energy efficiency certificates to meet their liability or set target. Depending on the scheme, one certificate usually represents one tonne of CO_2-e abated or one MWh of energy saved. Certificates are generated through approved energy efficiency activities. Certificates are usually created from energy reduction activities in both residential and/or commercial buildings. Generally, governments place an obligation on energy retailers to surrender an amount of certificates relevant to the retailer's market share. Retailers can generate their own credits through approved activities or can purchase certificates off accredited organisations. Retailers who fail to generate enough certificates can also purchase additional certificates off other retailers, thus enabling trading.

The Victorian Energy Efficiency scheme (VEES) is the only scheme that produces certificates based on carbon abated (i.e. 1 certificate = 1 tCO_2). As this scheme has been running since 2009, it existed in the presence of Australia's Front-End Carbon Pricing Mechanism, demonstrating that both a Front-End and End-User carbon trading scheme can exists simultaneously. This provides an excellent precedent and provides hope for future schemes.

In 2008, in a submission to a public consultation brief in New South Wales, development company Lend Lease recommended the adoption of a national Emissions Trading Scheme for the built environment (Lend Lease 2008). The proposal was significantly more comprehensive than then proposed state-based schemes as it also took into consideration the design and construction of buildings and the production of renewable energy onsite. It is believed that this proposed scheme be revisited and could be adapted further to include the additional areas identified in Chap. 12—the proposed framework for precinct-scale urban GHG emissions.

While the carbon trading scheme proposed at the time focused on individual buildings (not entire developments), there would be potential to extend it to wider urban development if suitable baselines could be established. These baselines could be established through the proposed carbon certification scheme for urban development. It would therefore function similar to the scheme proposed, in that developments that are below BAU or baseline GHG emissions, generate credits while developers above that level would need to purchase credits. The currency could be tradable carbon efficiency or intensity certificates rather than purely energy efficiency certificates, to allow for a range of other low-carbon initiatives. This is discussed further below.

13.5.9 Carbon Trading for the Built Environment

If a carbon trading scheme could be developed specifically for the built environment, it would reward developers for each tonne of carbon they abate below a given threshold, while requiring developments that emit over that threshold or baseline[19] to purchase carbon credits. This would require new carbon accounting tools for urban development in order to calculate and determine what the appropriate thresholds or baselines are. This is broadly known as baseline and credit scheme and is likely to be the most appropriate scheme for this sector. This type of scheme is likely to be more appropriate for the urban development sector than an energy efficiency certificate scheme, as the latter requires an entity to have an obligation to acquire certificates (i.e. an energy utility); it is unclear who would have the obligation when dealing with urban development.

Further research is needed to determine how such a scheme could fit into, or participate in, existing carbon schemes such as the Australian Direct Action Plan's Emission Reduction Fund (ERF).[20] The issue of additionality will still need to be dealt with, as well as how such a scheme would function in the presence of a RECs (or TGC) scheme and/or an energy efficiency scheme. As seen in the previous section, it is not impossible to have two carbon schemes operating concurrently, though care will be needed to streamline the schemes and avoid double counting.

Another issue that may affect developers participating in the ERF is the size of their projects. When deciding on projects to fund, the Australian Government will only consider large-scale abatement (i.e. over 2000tCO_2-e). Therefore, developments may need to be aggregated in order to provide large enough abatement to

---

19 Baselines could be set during two phases: over the construction period and during operation.

20 It should be noted that the ERF and the Direct Action Plan is not a trading scheme. It is purely crediting projects that demonstrate carbon abatement, using government money.

qualify for consideration. Mechanisms would need to be in place to ensure the ongoing monitoring of emissions from the aggregated sites.

Such a system could seek to be part of the national or global carbon trading system as long as appropriate certification of precinct developments was adopted following the full methodologies set out in this book. The generation and trading of these certificates could significantly enhance the economic viability and attractiveness of innovative low-carbon and carbon-neutral developments, particularly if they were part of the global trading of carbon credits.

13.6 Conclusion

While there are numerous benefits of pursuing low-carbon and carbon-neutral urban development for both developer and resident, such as reduced ongoing costs associated with living and working there, and marketing advantages for developers, there are also several barriers preventing or deterring developers from pursuing this form of development. This chapter reviewed some of the main barriers facing developers.

Various opportunities were then presented that could help to overcome some of these barriers. Government leadership was identified as a key factor in driving innovation in low-carbon development. In Australia, some governments were seen to be leading by example and demonstrating how to create low-carbon precincts. The city of Sydney and the Victorian Government's land development agency (Places Victoria) were provided as examples of government leadership in this chapter. Governments also have the ability to change legislation and regulations, which, in the energy industry, will be essential to allow and encourage new types of energy generation to enter the market and allow for new economic models that can facilitate this.

Creating platforms to share knowledge was also identified as important to progress the speed of transition towards low-carbon cities, as this allows developers to replicate successful measures and avoid less successful measures or failures. Platforms were also highlighted as a way to create baselines for the sector, allowing comparisons to be made between developments regardless of whether or not certification was sought. This could also provide the basis for a White Certificate Scheme or carbon trading for urban development.

Environmental upgrade agreements or EUAs at the precinct level were discussed as a mechanism that could assist developers dealing with the higher upfront costs associated with low-carbon infrastructure, by allowing some of those costs to be passed on to the residents through land rates. This idea was discussed further through the concept of a Business Improvement District (BID), though one that targeted specifically Green Infrastructure. Thus, a Greening, Regenerative and Improvement District (GRID) was proposed as the mechanism to facilitate the transition to Green Infrastructure in existing areas.

Energy Service Companies (ESCos) were acknowledged as a critical part of the new low-carbon economy that will likely play an important role in low-carbon developments in the future.

Low-carbon developments were identified in the literature as not yet being mainstream and therefore still requiring some form of assistance. As many of the barriers had financial implications, economic incentives were identified as a useful instrument to encourage greater uptake of the concept. Some of the incentives discussed, included the removal of stamp duty, land tax exemptions, a fast-track approvals process, Green Mortgages and various types of trading schemes (either carbon or energy efficiency) that could recognise and reward innovative developers and penalise poorly performing developments.

Finally, certification is proposed as a suitable mechanism for determining eligibility for the above-mentioned incentives, and a modified White Certificate (or energy efficiency) Scheme for the built environment (generating carbon certificates instead of energy efficiency) could be a possible mechanism for generating and trading global carbon credits.

References

Accenture. (No Date). Carbon disclosure project: The case for city disclosure: Carbon Disclosure Project.

ARUP & C40 Cities Climate Leadership Group. (2011). In *Climate action in Megacities: C40 cities baseline and opportunities*. Retrieved from http://www.arup.com/Publications/Climate_Action_in_Megacities.aspx.

Australian Sustainable Built Environment Council (ASBEC). (2008). *The second plank—Building a low carbon future with energy efficient buildings*. Retrieved from http://www.asbec.asn.au/files/ASBEC%20CCTG%20Second%20Plank%20Report%202.0_0.pdf.

Ball, J. (2002). Can ISO 14000 and eco-labelling turn the construction industry green? *Building and Environment, 37*, 421–428.

Blundell, L. (2012). Environmental upgrade agreements. In *The fifth estate*. Retrieved from http://www.thefifthestate.com.au/wp-content/uploads/2012/10/TFE_eBOOK_FINAL-41.pdf.

Bratt, C., Hallstedt, S., Robert, K. H., Broman, G., & Oldmark, J. (2011). Assessment of eco-labelling criteria development from a strategic sustainability perspective. *Journal of Cleaner Production, 19*, 1631–1638.

Brecard, D., Hlaimi, B., Lucas, S., Perraudeau, Y., & Salladarre, F. (2009). Determinants of demand for green products: An application to eco-label demand for fish in Europe. *Ecological Economics, 69*, 115–125.

Brisbane City Council. (2013). Kenmore *Suburban Centre Improvement Project* (*SCIP*). Retrieved from http://www.brisbane.qld.gov.au/planning-building/planning-guidelines-tools/urban-design-projects/suburban-centre-improvement-projects.

Bunning, J., Beattie, C., Rauland, V., & Newman, P. (2013). Low-carbon sustainable precincts: An Australian perspective. *Sustainability, 5*(6), 2305–2326.

Climate Change Authority. (2012). *Renewable energy target review: Discussion paper*. Retrieved from http://climatechangeauthority.gov.au/ret.

Clean Energy Future. (2013). *How Australia's carbon price is working one year on*. Retrieved from https://www.thepmr.org/system/files/documents/0045+Carbon+Price+One+Year+On+18_06+d_1_0.pdf.

Cones, J. D., & Hayes, S. C. (1980). *Environmental problems/behavioural solutions*. Monterey, California: Brooks/Cole Publishing Company.

CSIRO. (2009). Intelligent grid report. In *A value proposition for distributed energy in Australia*. Commonwealth Scientific and Industrial Research Organisation.

Currie, K. M. (2013). A regulatory adjustment process for the determination of the optimal percentage requirement in an electricity market with tradeable green certificates. *Energy Policy, 62*, 1053–1057.

Deneen, S. (n.d.). Green mortgages. In *The daily green*. Retrieved from http://www.thedailygreen.com/living-green/definitions/green-mortgages.

Department of Climate Change and Energy Efficiency (DCCEE). (2010). *Report of the Prime Minister's task group on energy efficiency*. Retrieved from http://www.industry.gov.au/Energy/EnergyEfficiency/Documents/energy-efficiency/workshops-report.pdf.

Department of Industry, Innovation, Climate Change, Science, Research and Tertiary Education (DIICCSRTE). (2013). *Australian national greenhouse accounts: National inventory report 2011*. Retrieved from http://www.climatechange.gov.au/sites/climatechange/files/documents/05_2013/AUS_NIR_2011_Vol1.pdf.

Fudge, S., & Peters, M. (Eds.). (2014). *International approaches to behaviour change: The global challenge to encouraging sustainable lifestyles*. London: Edward Elgar.

Garnaut, R. (2008). *The Garnaut climate change review—Final report*. Melbourne: Cambridge University Press.

GBNYC. (2012). *New York state passes real property tax exemption legislation for green buildings*. Retrieved from http://www.greenbuildingsnyc.com/2012/07/30/new-york-state-passes-real-property-tax-exemption-legislation-for-green-buildings/.

Glass, J., Dainty, A. R. J., & Gibb, A. G. F. (2008). New build: Materials, techniques, skills and innovation. *Energy Policy, 36*, 4534–4538.

Government of Western Australia. (2012). *2012–13 Land tax*. Brochure. Retrieved from http://www.finance.wa.gov.au/cms/uploadedFiles/_State_Revenue/Land_Tax/Land_Tax_Information_Brochure.pdf?.

Hoyt, L. (2004). Collecting private funds for safer public spaces: an empirical examination of the business improvement district concept. *Environment and Planning B: Planning and Design, 31*, 367–380.

Hwang, B. G., & Ng, W. J. (2013). Project management knowledge and skills for green construction: Overcoming challenges. *International Journal of Project Management, 31*, 272–284.

International Labour Office (ILO). (2011). *Skills and occupational needs in green building*. Geneva: European Commission.

Jones, P., & Patterson, J. (2007). The development of a practical evaluation tool for urban sustainability. *Indoor and Built Environment, 16*(3), 255–272.

Kennedy, C., & Corfee-Morlot, J. (2013). Past performance and future needs for low carbon climate resilient infrastructure—An investment perspective. *Energy Policy, 59*, 773–783.

Lehmann, S. (2013). Low carbon construction systems using prefabricated engineered solid wood panels for urban infill to significantly reduce greenhouse gas emissions. *Sustainable Cities and Society, 6*, 57–67.

Lehmann, S. (2015). *Low carbon cities: Transforming urban systems*. New York: Routledge.

Lend Lease. (2008). *Submission by lend lease corporation in response to NSW energy efficiency trading scheme—Discussion paper*. Retrieved from http://www.energy.nsw.gov.au/sustainable/efficiency/scheme/submissions-2008/sustain_neet_lend_lease.pdf.

Levy, P. R. (2001). Paying for the public life. *Economic Development Quarterly, 15*, 124–131.

McKenzie-Mohr, D. (2007). *Community based social marketing*. Retrieved from http://www.cbsm.com.

Mobbs, M. (2010, September 28). Bathurst burr: Do Australian governments support sustainable development? Errrrr… maybe not so much…. In *The fifth estate*. Retrieved from http://www.thefifthestate.com.au/archives/16251/.

Munoz, S. S. (2007, September 12). Going green to save some green: Lenders push mortgages with discounts and credits for energy-efficient upgrades. *The Wall Street Journal.* Retrieved from http://online.wsj.com/public/article/SB118955748175824511.html.

Moreland Energy Foundation, Net Balance, & Green Spark Consulting. (2011). *Business models for enabling sustainable precincts.* Sustainability Victoria. Retrieved from http://www.mefl.com.au/online-library/research-and-project-reports-1/business-models-for-enabling-sustainable-precincts-1/256-business-mdels-for-enabling-sustainable-precincts-report-1/file.html.

Newman, P., & Kenworthy, J. (2015). *The end of automobile dependence: How cities are moving beyond car-based planning.* Washington DC: Island Press.

Newman, P., Beatley, T., & Boyer, H. (2009). *Resilient cities: Responding to peak oil and climate change.* Washington DC: Island Press.

Newman, P., Rauland, V. & Holden, D. (2011). *Creating resilient cities: How a new generation of tools can assist local governments in achieving their carbon abatement goals.* Paper presented at the State of Australian Cities National Conference, Nov 29–Dec 2 2011, Melbourne, Victoria

Newton, P. (Ed.). (2008). *Transitions: Pathways towards sustainable urban development in Australia.* Melbourne: CSIRO Publishing.

O'Leary, T. (2012). *Residential energy efficiency and mandatory disclosure practices.* Paper presented at the Eighteenth Annual Pacific-Rim Real Estate Society Conference, January 15–18, 2012, Adelaide, SA.

Organisation for Economic Co-operation and Development (OECD). (2002). *Towards sustainable household consumption?.* Paris: Trends and Policies in OECD Countries.

Pacifici, C. (2011). Greening the real estate industry. In *2011 Update: Identifying the best way to build capacity within the real estate profession to communicate sustainability measures to sellers, buyers, landlords and tenants.* Retrieved from http://greengurus.com.au/wp-content/uploads/2011/03/Greening-the-Real-Estate-Industry_2011Report.pdf.

Partners in Project Green. (n.d.). Pearson eco-business zone policy toolkit: Primers. In *Fast tracking eco-business zone (EBZ) application: How to get started. GTAA partners in project green—A Pearson Eco-Business Zone.* Retrieved from http://ogph.ca/admin/category-image/4-%20Primers_Fast%20Tracking%20EBZ%20Apps.pdf.

Pettit, C. J., Glackin, S., Trubka, R., Ngo, T., Lade, O., Newton, P., & Newman, P. (2014). A Co-design prototyping approach for building a precinct planning tool. *ISPRS Annals of the Photogrammetry, Remote Sensing and Spatial Information Sciences, 11*(2), 47–53.

Planning Institute Australia (PIA). (2011). *Favouring green applauded by planners.* Media release. Queensland Division: Planning Institute of Australia. Retrieved from http://www.planning.org.au/documents/item/3053.

Queensland Government. (2011). Media statement. In *Opening the green door for sustainable development projects.* Retrieved from http://statements.qld.gov.au/Statement/Id/75660.

Rauland, V. (2013). *Decarbonising cities: Certifying carbon reduction in urban development.* Curtin University Sustainability Policy (CUSP) Institute: Curtin University. (Doctoral Dissertation).

Rauland, V. (2014). *How to fund low carbon urban development.* Retrieved from http://www.thefifthestate.com.au/archives/60778/.

Rauland, V., & Thomson, G. (2015). Mainstreaming Sustainable Precincts: sharing experiences. *In Fifth Estate eBook: Creating Sustainable Precincts.* 88–97. Retrieved from http://www.thefifthestate.com.au/wp-content/uploads/2014/12/Precincts_ebook_final-double-page.pdf.

RedEnergy. (2014). *Energy efficiency schemes.* Retrieved from http://www.redenergy.com.au/page.html?energy-efficiency-schemes.

Revitalising Central Dandenong. (n.d.). Precinct energy project. Retrieved from http://www.revitalisingcentraldandenong.com/cs/Satellite?c=VPage&cid=1339118124865&pagename=RCD/Layout.

Roseland, M. (1997). *Eco-city dimensions: Healthy communities, healthy planet.* Gabriola Island, Canada: New Society Publishers.

Sayce, S., Ellison, L., & Parnell, P. (2007). Understanding investment drivers for UK sustainable property. *Building Research and Information, 35*(6), 629–643.

Sedlacek, S., & Maier, G. (2012). Can green building councils serve as third party governance institutions? An economic and institutional analysis. *Energy Policy, 49*, 479–487.

Shaheen, S. A., & Cohen, A. P. (2013). Carsharing and personal vehicle services: Worldwide market developments and emerging trends. *International Journal of Sustainable Transportation., 7*(1), 5–34.

Sheehan, P., Cheng, E. J., English, F. A., & Sun, F. A. (2014). China's response to the air pollution shock. *Nature Climate Change, 4*, 306–309.

Sinclair, H. M. W. (2008). *Carbon neutral cities: The need for more sustainable development options in response to climate change.* Planning and Urban Development Program, University of New South Wales. (PLAN 4132: Thesis Project).

Szatow, A., Quezada, G., & Lilley, B. (2012). New light on an old problem: Reflections on barriers and enablers of distributed energy. *Energy Policy, 43*, 1–5.

Toderian B (2013) Six tips for city hall leadership, *Planetizen,* August 8.

Thomson, G., & Rauland, V. (2014). GRID: A new governance mechanism for financing eco-infrastructure at the district scale.In 7th *International urban design conference—Designing productive cities, September 1–3 2014.* Adelaide, SA: Urban Design Australia.

Thomson, G., Matan, A. & Newman, P. (2013). *A Review of International Low Carbion Precincts to identify Pathways for Mainstreaming Sustainable Urbanism in Australia.* 7th State of Australian Cities Conference, Sydney, NSW.

Toze, S. (2006). Water reuse and health risks— real versus perceived. *Desalination, 187*, 41–51.

Williams, J. (2012). *Zero carbon homes: A road map.* New York: Earthscan.

Wills, R. (2011). *Smart, smarter, smartest 'Zigbee preparing for the next wave of home innovation.*' Retrieved from http://www.thebuilderschoice.com.au.

World Business Council on Sustainable Development (WBCSD). (2009). *Transforming the market: Energy efficiency in buildings.* Retrieved from http://www.wbcsd.org/transformingthemarketeeb.aspx.

Zaw, Y. (2012). Share scheme to cut use as Freo grows. In *The west. Web page*. Retrieved from http://au.news.yahoo.com/thewest/a/-/breaking/13320093/share-scheme-to-cut-use-as-freo-grows/.

Zuo, J., Read, B., Pullen, S., & Shi, Q. (2012). Achieving carbon neutrality in commercial building developments—Perceptions of the construction industry. *Habitat International, 36*, 278–286.

Chapter 14
Final Thoughts

14.1 Introduction

This book has examined the significant potential of cities and urban development to reduce global greenhouse gas emissions, and the question of whether the carbon reductions achieved can be mainstreamed by recognising them under formal certification system. While the book has shown that considerable emission abatement is possible from this sector, there are several issues that need to be worked through in order to certify reductions, many of which relate to governance. It also identified the need to reward progressive developers.

Historically, the process of tackling carbon has focused on the Front-End of the economy and has adopted a top-down approach using measures such as emission trading schemes and carbon taxes. This has resulted in rigorous carbon accounting methodologies being developed for power companies and large industry. As this has generally been seen as sufficient in addressing emissions, few specific economic or regulatory measures and schemes for reducing carbon have been developed at the other end of the economy, which includes urban development and the built environment. The Front-End approach has tended to relegate End-User innovations in low-carbon to a lower level of activity that should somehow flow through once a carbon price is set. As a result, limited carbon accounting processes, methodologies and frameworks currently exist to measure emissions from the built environment sector, and this prevents accurate reporting and subsequent recognition of reductions. It has also meant that policy innovation in cities has been somewhat sporadic.

However, as awareness of the carbon reduction potential from the built environment has grown—a result of numerous low-carbon and carbon-neutral urban development demonstration projects around the world—the need for a consistent way to measure emissions, and acknowledge and facilitate the reduction, has become evident. The reality is that cities change quite rapidly and can indeed be the catalyst for radical change of the kind that the world is requiring (IPCC 2014).

V. Rauland and P. Newman, *Decarbonising Cities*, Green Energy and Technology, DOI 10.1007/978-3-319-15506-7_14

Chinese cities have demonstrated this in recent years (Sheehan et al. 2014; Newman and Kenworthy 2015).

We need to find a way to integrate both Front-End and End-User approaches when targeting carbon abatement in the economy, ideally through a trading mechanism.

Identifying new ways of promoting and incentivising low-carbon and carbon-neutral urban development in the short term is a new challenge for government, industry and community. Providing urban development with the ability to generate and trade carbon credits could potentially help to mainstream the decarbonising of cities.

The following sections address the main issues posed at the beginning of this book.

14.2 The Carbon Reduction Potential of Urban Development

The GHG contribution of urban areas is considerable and is estimated to range between 40 and 80 % of global emissions, depending on definitions. The built environment is a large component of this and has long been recognised for its potential to reduce emissions. Buildings, which are large consumers of energy, have been the primary focus of emission reduction over the years. As a result, a plethora of building assessment tools now exists, which has led to a thriving green building industry.

However, in recent years, a shift in focus has occurred from the building to the wider precinct scale, as building designers, developers, local governments and urban planners have identified the greater benefits, efficiencies and synergies that can be achieved when measures, such as energy generation, are implemented at the precinct scale. This scale is also particularly important for addressing transport GHGs, which are often not addressed at the building scale.

Various demonstration projects around the world have trialled and tested different technologies and alternative ways of designing and constructing urban precinct development, which include new and innovative ways of managing resources locally and reducing car dependence. These have demonstrated significant carbon reductions. However, these developments remain 'demonstration' or 'niche' projects that require various forms of assistance (Thomson et al. 2013). If these types of projects are not supported, encouraged and recognised for their carbon reduction achievement, this growing social movement is likely to collapse.

14.3 Certifying Low-Carbon Urban Development

Carbon certification is identified as a way of acknowledging and rewarding progressive urban developments that can demonstrate genuine carbon reductions. Carbon certification is now a global industry, underpinned by strict standards and regulations designed to prevent false claims being made and thus significantly helping to increase the credibility of claims.

The carbon certification industry primarily targets End-Users within the voluntary sector, i.e. those whose emissions are not captured under mandatory carbon reduction activities, and it has focused on businesses and organisations. The architecture of existing certification schemes has therefore been designed to deal with this relatively straightforward sector (i.e. businesses operational emissions). As a result, various issues arise when carbon certification is applied to precinct-scale urban development and the built environment, as this sector has a different set of conditions and complexities associated with it.

This book has shown that several key factors need to be considered when designing a certification scheme for urban development. These are addressed below.

14.3.1 Core Elements of Carbon Certification for Urban Development

Creating a carbon certification scheme specifically for the development of urban precincts ideally needs to consider five core elements as follows: a standardised framework for calculating the carbon associated with six key areas of urban development; online tools to help with the carbon calculations in the design phase, assessment and ongoing monitoring of operational emissions; the identification or creation of a local body responsible for managing (internally) the ongoing certification process post-construction; the adoption of engagement and behaviour change programs for people moving into the development to encourage the behaviours required to keep operational emissions low; and new regulations to establish this whole process in the planning system and to enable onsite emission reductions. These core elements are illustrated in Fig. 14.1.

Fig. 14.1 Core elements for low-carbon certification for urban development

The carbon accounting framework proposed in this book is the first attempt to provide a comprehensive, standardised approach to calculating emissions at the precinct level in order to create a global GHG standard for the urban development sector.

14.4 Barriers and Opportunities—Increasing Uptake of Low-Carbon Urban Development

Several barriers have been identified in this book that are preventing greater adoption of the concept of low-carbon and carbon-neutral urban development. These include higher upfront capital costs associated with newer technologies and designs, first mover disadvantage, the credibility of carbon claims, information barriers, split incentives, lock-in, policy and pricing uncertainty, dealing with multiple stakeholders and issues with utilities.

Various opportunities were also identified that could assist in overcoming many of the barriers, thus helping to facilitate a greater uptake of low-carbon development. As many of the barriers resulted in additional development costs, economic incentives were seen as an important measure to encourage developers to pursue this form of development in the short-term, until they become mainstream. Carbon certification was identified as a useful way of determining the eligibility of developments for receiving the financial incentives discussed.

This book has proposed the idea of creating a new carbon trading scheme for the built environment (including precinct-scale development), which builds on the successes of existing energy efficiency or white certificate schemes. Such a scheme could use the carbon abated from a range of low-carbon initiatives as the tradable commodity instead of energy efficiency activities, to create an End-User carbon trading scheme. This could be applied in areas where substantial growth in cities is happening or expected (especially in emerging nations) and where a Front-End carbon price has not yet been established. It can also be worked out as a parallel or integrated mechanism to assist with Front-End approaches, in countries where carbon pricing is already established. Such a carbon credit mechanism would require all the other elements outlined in this book regarding accounting, assessment and governance to be further developed. It would enable cities to be better planned in the process of achieving the carbon credits, as the multiple co-benefits are tangible.

Table 14.1 highlights the various barriers and opportunities and demonstrates the interconnections between them and how the barriers can be addressed.

The core elements for certification identified in this book, combined with the opportunities presented above, could provide a powerful way of mainstreaming the decarbonisation process in cities. This is demonstrated in Fig. 14.2.

Table 14.1 Barriers and opportunities for low-carbon development

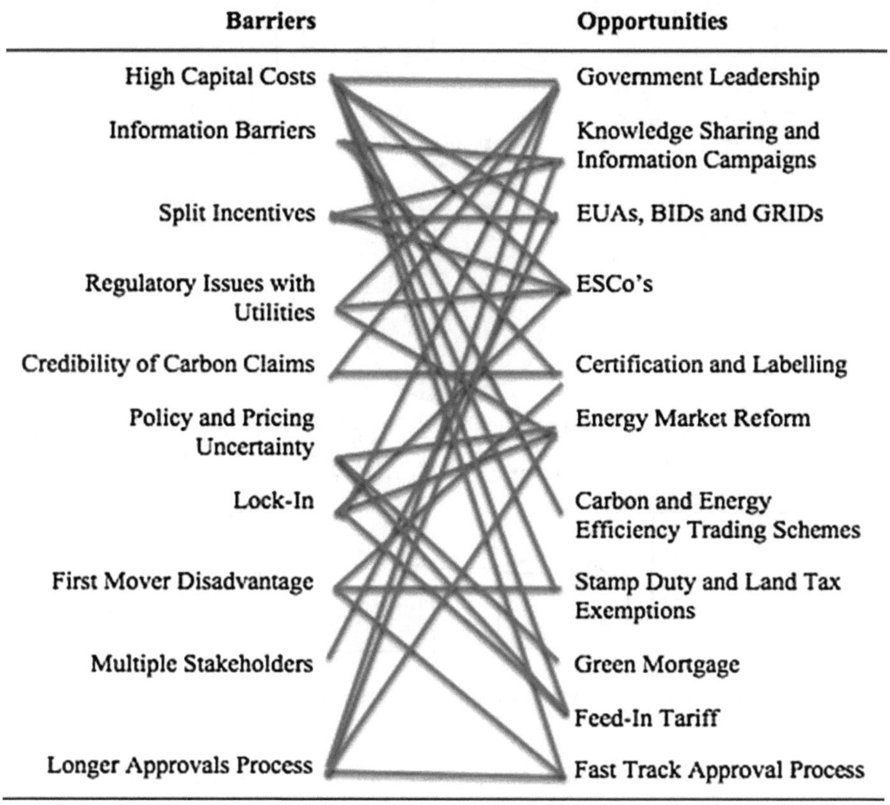

14.5 Next...

To mainstream low-carbon urban development as a way of decarbonising cities, the opportunities discussed in this book need to be taken and trialled, ideally as a package. The new governance systems discussed need more attention than the development of technologies and designs for decarbonising cities, most of which already exist. In particular, a new set of tools needs to be developed, which include streamlined, online carbon accounting and assessment tools that are affordable and easy for developers to use, as well as for city or local governments to use in their planning assessment and programs for decarbonising. Turning the proposed carbon accounting framework into an ISO standard for carbon accounting for urban development at the precinct scale will also be critical to ensure greater consistency and comparability between developments in various countries.

New tools will be able to unlock many more innovations in the soft systems (i.e. governance) as well as the hardware (i.e. technologies and infrastructure) needed for low-carbon or zero-carbon cities. Though, at the end of the day, cities are about

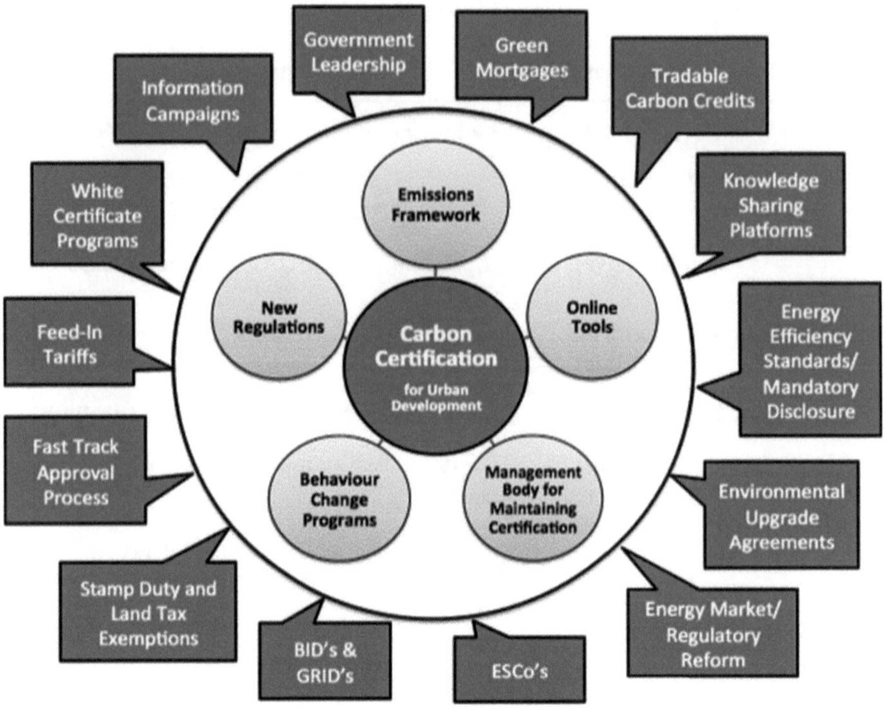

Fig. 14.2 Opportunities to assist with mainstreaming the concept of low-carbon development

people, and a plethora of cultural and political options can be realised if people can recognise the benefits that can flow from decarbonising cities. This requires greater access to information.

In terms of new structures, there are two important developments that are suggested to be pursued and trialled:

1. The application of the concept of the 'Greening, Regenerative, and Improvement District' or GRID (based on the business improvement district) which has significant potential to help manage multiple stakeholders and the costs associated with implementing green infrastructure at the local precinct level, and
2. The increased adoption of ESCos and small-scale utilities (including water) to run low-carbon services locally, which is likely to require changes to regulation, and support from centralised utilities and government departments managing energy, water and waste legislation.

These innovations will be needed in all cities with their cultural variations that can enable us all to learn what works best. Jane Jacobs (1984) suggests that wealth is created by innovations in cities copying each other and improving on the innovations in the process. Certainly, decarbonising innovations will be a major source of wealth in the future and will be unlocked by such mechanisms as these.

Finally, the question of sending a price signal for precinct-scale development is important to consider, especially in those parts of the world where there is substantial growth in cities and where a Front-End carbon price has not yet been implemented. To be mainstreamed, low-carbon urban innovations need to have a market in these times of transition. With a well-developed carbon accounting framework and certification process in place for precinct-scale development, it is possible to take the next step and create a mechanism for trading carbon certificates within the built environment.

Demonstrations of End-User carbon trading schemes already exist and include Tokyo's carbon trading scheme targeting commercial buildings, as well as various building energy efficiency schemes, which allow trading. Australia's Direct Action Plan (DAP) is another policy that has the potential to enable carbon credits to be generated from the built environment, though methodologies still need to be developed (particularly for precinct-scale development) and trading is not currently possible.

Further work is needed to determine how a carbon trading system targeting the built environment would be designed. Such a scheme could borrow from successful examples of energy efficiency schemes around the world. How this kind of End-User carbon credit system can be fitted into the global Front-End carbon credit system remains to be seen, particularly in relation to dealing with double counting at the national level, but it is worth pursuing. Such a system has the potential to create a significant financial incentive to decarbonise cities.

Mainstreaming low-carbon urban development is a critical agenda. This book has shown that the technologies are available, the urban design principles for decarbonisation are well understood, but the governance systems to facilitate and manage precinct-scale development are still largely lacking. If we can develop such systems, processes and mechanisms that target the built environment and reward positive behaviour, we could begin to radically decarbonise our cities. It is not beyond us.

References

Intergovernmental Panel on Climate Change [IPCC]. (2014). *IPCC Working Group 3: Mitigation of climate change, assessment report 5 of Working Group 3*. Geneva: Intergovernmental Panel on Climate Change.

Jacobs, J. (1984). *Cities and the Wealth of Nations*. New York: Random House.

Newman, P., & Kenworthy, J. (2015). *The end of automobile dependence: How cities are moving beyond car-based planning*. Washington DC: Island Press.

Sheehan, P., Cheng, E. J., English, F. A., & Sun, F. A. (2014). China's response to the air pollution shock. *Nature Climate Change, 4*, 306–309.

Thomson, G., Matan, A. & Newman, P. (2013). A review of international low carbon precincts to identify pathways for mainstreaming sustainable urbanism in Australia. 7th *State of Australian Cities Conference, Sydney, NSW*.

Appendix A
Key Considerations for Carbon Footprint Definitions

Gases

The first clarification required when defining a carbon footprint relates to the types of gases covered. In their paper 'A Definition of a Carbon Footprint', Wiedmann and Minx (2008) question whether a carbon footprint, to be represented accurately, should include only gases that contain the carbon molecule (i.e. CO_2, CH_2, CH_4). However, this would mean disregarding other potentially potent GHGs such as N_2O. The authors, however, confine this even further by proposing their own definition of a carbon footprint as consisting exclusively of carbon dioxide, largely for simplicity. They suggest that for calculations involving multiple greenhouse gases, a more appropriate title would be 'climate footprint' (Wiedmann and Minx 2008). Nevertheless, most carbon footprint calculations and definitions within the public domain consider a range of greenhouse gas emissions, predominantly the six GHGs identified and covered under the Kyoto Protocol (Table A.1).

As each GHG has a different effect on the atmosphere, a specific global warming potential (GWP) is given to it. The GWP is determined based on the impact each gas has on the atmosphere over a 100-year period, compared to a reference gas. Carbon dioxide has been elected as the reference gas and thus has a GWP of one (East 2008; Garnaut 2008). This ensures that all gases are comparable and additive (Galli et al. 2011). The GWP for the six GHGs covered under the Kyoto Protocol, which are the most common choice of gases for GHG reporting, are listed in Table 5.

By standardising the units, more consistent and straightforward emission calculations can be made using a single calibrated unit. These gases are thus represented as carbon dioxide equivalents (CO_2-e), and this is the dominant unit used in carbon calculations.

This book proposes that a carbon footprint (and any subsequent carbon claim) for the built environment and precinct-scale development should contain the six main GHGs as outlined in the Kyoto protocol. There are several reasons for this.

First, as demonstrated in Chaps. 3 and 4, emissions attributable to the built environment stem from a variety of different emission sources, and these produce various types of emissions. Therefore, in order to obtain a comprehensive emissions

V. Rauland and P. Newman, *Decarbonising Cities*, Green Energy and Technology, DOI 10.1007/978-3-319-15506-7

Table A.1 The six Kyoto greenhouse gases and their global warming potential

Kyoto greenhouse gas	Chemical formula	Global warming potential	Major sources (examples)
Carbon dioxide	CO_2	1	Fossil fuel combustion, land use change, cement manufacturing
Methane	CH_4	21	Natural gas, coal mines, anaerobic decomposition of organic waste
Nitrous oxide	N_2O	310	Fertilisers, agriculture, combustion
Perfluorocarbons	(7 types)	7,000	Electronics, aluminium production
Hydrofluorocarbons	(13 types)	11,700	Refrigerants
Sulfur hexafluoride	SF6	23,900	High voltage switchgear

profile, as many of these sources should be included as practicable. For example, many of the emissions reduction opportunities existing within the built environment often involve switching gases (e.g. from coal to gas), each with a varying GWP, or managing resources more efficiently (e.g. capturing the methane from organic waste to use as energy), thereby demonstrating the need for a range of gases to be incorporated into carbon calculations.

Second, by ensuring that the carbon accounting process aligns with already established protocols and mandatory emission reporting frameworks (such as the Kyoto Protocol, Australia's NGERS Framework and the Government's NCOS Carbon Neutral Program), participants who are already required to comply with or are voluntarily using an existing scheme can benefit by using the same information and accounting methods and tools. Organisations, local councils and developers who voluntarily undertake a carbon audit or footprint that is based on methods used in existing regulatory frameworks (i.e. NGERS) help prepare and future-proof themselves for a time when changes to regulations occur.[1]

Measurement Units

The concept of a carbon footprint originated from the ecological footprint model developed by Rees and Wackernagle (1996), which was measured in global hectares. However, in order to be consistent with other types of GHG or carbon measurement and reporting processes, the measurement unit chosen for carbon footprinting has been in weight or mass (i.e. kg or tonnes of CO_2 or CO_2-e). Therefore, as Hammond (2007) suggests, a more accurate term would be 'carbon weight' as opposed to 'footprint', which intuitively makes one think of land area. However, the term 'footprint' is likely to remain the popular and preferred word

[1] As stipulated in the National Greenhouse Gas and Energy Reporting Scheme guidelines, the reporting threshold will reduce over time to capture additional sectors and organisations.

despite its unusual measurement unit, as it is much more visual and easier to understand and hence, more meaningful than the idea of carbon as weight. Another benefit of using the term footprint is the underlying connotation of a footprint representing a legacy, i.e. the long-term impact carbon has on the earth and atmosphere.

Carbon footprints are primarily measured in metric tonnes as they usually represent significant quantities of gas. However, they are generally converted into kilograms during the calculation process as emission factors are primarily provided in this smaller unit (DCCEE 2011b). If the measurement is being used to describe small quantities, such as the carbon footprint of a product, the measurement unit is usually kept in kilograms (Bolwig and Gibbon 2009).

A recent carbon footprint undertaken for a local high school (Box 1) demonstrated that many of the emission factors provided in publicly available Australian documents were not suitable for small-scale operations and thus made emissions calculations difficult (Rauland and Hall 2012). This is most likely due to the fact that most of the GHG emissions factors, calculations and reporting guidelines are geared towards large emission-intensive industries such as energy generation, refining, mining and other energy-intensive materials-based manufacturing. These companies and activities are generally captured under mandatory reporting (i.e. NGERS) and the recently introduced carbon pricing legislation. This lack of data available for smaller scale activities, such as the appropriate emissions factors, may prove to be an issue for land developers wishing to report emissions for their developments.

Accounting Methodology

As discussed earlier, GHG accounting methodologies vary depending on the entity being assessed. Organisations, services and personal carbon footprints generally follow an inventory style GHG accounting approach, which focus primarily on the operational emissions. This approach is largely based on The GHG Protocol: A Corporate Accounting and Reporting Standard (WRI and WBCSD 2004), which also underpins the various Australian and international GHG accounting Standards, e.g. ISO 14064-1 (ISO 2006) and AS/NZS I46064-1 (Standards Australia 2006). More information about the boundaries and Scopes of emissions associated with this GHG accounting methodology are provided below.

Cities and local governments tend to adopt this operational inventory approach when measuring their emissions, as it is far less complicated than undertaking a full life cycle GHG assessment (Hoornweg et al. 2011; ICLEI 2010). This is particularly important for cities, which are already such large and complex systems (Dhakal 2010; Kennedy and Sgouridis 2011).

A life cycle inventory analysis is the other predominant GHG accounting methodology, and involves examining emissions along the entire supply chain, from production and acquisition of raw materials, to transport, manufacturing, storage and ultimate disposal. This approach is most often applied to products (DCCEE 2011a; Matthews et al. 2008).

It is argued that precinct-scale development would benefit from using both approaches; a life cycle approach in quantifying emissions would be needed due to the significant emissions attributed to factors such as materials, construction processes and the resource management systems underpinning a development. At the same time the design of buildings, density and mix of the development will largely determine the operational GHGs, including transport GHG outcomes and the building GHGs. Understanding the entire footprint of a development allows developers to identify the easiest carbon abatement opportunities (i.e. switching materials, processes or technologies as well as the density and mix issues). Developers are therefore able to make the greatest reductions to their carbon footprint through more informed purchasing and planning decisions (Matthews et al. 2008).

There are several methodologies and ways to undertake a Life Cycle Analysis, which can lead to very different GHG outcomes (Nässén et al. 2007). The main approaches include Process Analysis (PA), Economic and/or Environmental Input-Output (EIO) Analysis, or a Hybrid EIO-LCA (Berners-Lee et al. 2011; Matthews et al. 2008). Process Analysis is the most commonly applied method for conducting a Life Cycle Analysis and involves systematically identifying and quantifying all emissions along the supply chain, beginning with direct emissions before tracking all indirect emissions. While this process provides superior specificity in terms of data collection and overall numbers, the comprehensiveness of the process (i.e. including each section of the supply chain that has its own additional set of inputs/outputs) means that the analysis is very time consuming as it could theoretically continue indefinitely (Berners-Lee et al. 2011).

Economic Input-Output Analysis is another widely used approach, which uses large-scale, aggregate data based on industry sectors within an economy. The information is gathered using financial expenditure, i.e. by examining each sector's interdependence based on their financial transactions.[2] If the EIO data for each sector is available and a sufficient accounting system is in place for the product being examined, this method can prove relatively easy compared to the PA approach (Berners-Lee et al. 2011). Nevertheless, such large data sets mean averages must be used, which limits the overall accuracy of the approach.

The EIO is often referred to as a top-down approach while the PA is seen as a bottom-up approach. A hybrid of these two models has also been suggested as a way of overcoming the limitations and weaknesses of both approaches and benefitting from their strengths. A hybrid EIO-LCA involves using the basic EIO approach but, instead of utilising the broad sector data, it is "augmented with impact data for specific goods, services, and organisations" (Matthews et al. 2008: 5840).

[2] By analyzing the financial transactions of one industry, additional industries and sectors can be identified that may need to be included in the analysis. This is essentially identifying supply chain emissions.

A typical life cycle assessment approach may not necessarily be the most appropriate choice for precinct development, given the complexity associated with assessing multiple emission sources within a precincts boundary. It is therefore suggested that a hybrid approach would be the most suitable option for development, which effectively follows the Hybrid EIO-LCA described above. Adopting a hybrid approach for calculating the emissions associated with construction is also recommended in the Federal Government's National Carbon Offset Standard Carbon Neutral Program (DCCEE 2011a).

There are several tools that can assist in undertaking the calculation process. These were discussed in Chap. 9.

Boundaries

Determining the boundaries of a carbon footprint or GHG inventory is one of the first steps in any carbon accounting process and the absence of agreed boundaries is one of the key causes of inconsistency among carbon precinct claims. Boundaries can be set at various levels. If the focus is on corporate carbon accounting, two levels of boundaries need to be established—the organisational boundary and the operational boundary. If a life cycle carbon analysis is the chosen methodology, boundaries are not generally associated with organisational or operational control but with the supply chain. As highlighted above, determining clear boundaries associated with the life cycle approach is critical as boundaries are essentially infinite (Matthews et al. 2008).

An organisational boundary is determined based on full ownership or financial control of the organisation. The GHG Protocol: A Corporate Accounting and Reporting Standard (WRI and WBCSD 2004) outlines two approaches for establishing organisational boundaries, the 'Equity share approach' and the 'Control approach'. The equity share approach requires corporations who have equity in other organisations, subsidiaries, partner or joint ventures to report emissions from those companies depending on the percentage they own. The control approach only requires corporations to report on emissions from organisations, subsidiaries, partner or joint ventures that they have financial or operational control over (WRI and WBCSD 2004).

The organisational boundary is not applicable to precinct-scale land development if the aim is to achieve carbon neutrality. While a developer will be responsible for the majority of the emissions associated with the construction of a development, there are several additional emission sources that would not be captured under this boundary approach, such as the emissions generated from the utilities supplying resources to the area (e.g. energy, water and managing waste).

The operational boundaries outlined in The GHG Protocol (WRI and WBCSD 2004) are those emissions associated with the operations of the company/corporation, once they have identified their organisational control. These emissions are generally identified using 'Scopes', which are discussed in detail below. Taken together, the organisational and operational boundaries represent the overall system boundary of an organisation.

Again, in relation to precinct development, neither of these boundaries are likely to be relevant, as a precinct, particularly after construction, is likely to consist of multiple owners, stakeholders and organisations, which all ultimately contribute to the emissions profile of the precinct.

An additional boundary that is often discussed in relation to local governments and cities is the geographic and geopolitical boundary, which consists of the official jurisdictional boundaries of the physical land area associated with a local government or city (ICLEI 2009). Kennedy and Sgouridis (2011) also discuss the issue of geographic boundaries in relation to carbon accounting in cities, stating that:

> A city is a dynamic and complex system, defined in part by its geographic boundaries, but also by its interconnections with a much broader region through exchanges of materials, energy, and information. This interconnectedness complicates the task of determining which emissions should be included in a city's carbon balance. (p: 5261)

The boundary issue often stimulates discussion around GHG attribution (i.e. whether emissions should be accounted for at the place of production or consumption), which was discussed in Chap. 3.

With respect to transport, a precinct can only have limited responsibility from an organisational perspective, but from a local government's perspective it is very important. Thus, in order to achieve compliance with local governments carbon reduction goals, a developer may have to accept that there is a role for them in terms of urban design, density and mix, which will have an impact on the carbon outcomes. While the provision of infrastructure for public transport has historically been a state or federal responsibility, in the future it is likely to require some kind of partnership approach and will cross boundaries beyond most precinct plans.

Scopes

Emission sources are broadly split into two categories, direct and indirect emissions. Direct emissions are those that occur on-site or through activities that are under the full control of an organisation. Indirect emissions are those that are produced by a separate organisation/entity (e.g. electricity production or emissions from aircrafts) but still occur as a result of demand by the first organisation/product (WRI and WBCSD 2004).

The concept of Scopes for emissions was introduced as a way of dealing with direct and indirect emission sources in order to avoid issues associated with double counting. Double counting occurs when the same emission sources are counted twice by different organisations, and is principally only an issue if the emissions are being reported under a mandatory reporting or emissions trading scheme (WRI and WBCSD 2004). It is less consequential if emissions are being voluntarily reported,

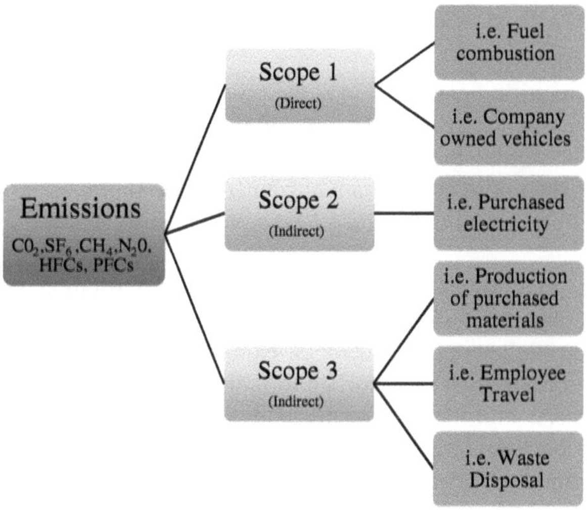

Fig. A.1 Overview of scopes and emissions. *Source* adapted from WRI and WBCSD (2004)

and indeed, it is argued that the more indirect emissions that are included under voluntary reporting, the better (DCCEE 2011a).[3]

Emissions are thus defined as Scope 1, Scope 2 and Scope 3 emissions as depicted in Fig. A.1. Scope 1 emissions are direct emissions from sources within an organisation's direct control. Scope 2 are indirect emissions from electricity production and Scope 3 emissions are indirect emissions from all other activities related to the organisation. Mandatory reporting in Australia (i.e. under the National Greenhouse and Energy Reporting Act) requires covered sectors and organisations to report only their Scope 1 and 2 emissions. This is because the majority of the emissions produced in large, emission intensive industrial companies come from Scopes 1 and 2 sources (Fig. A.1).

For the goods and services sectors, however, Scopes 1 and 2 can represent less than 25 % of their total emissions (Matthews et al. 2008). Therefore, if companies voluntarily choose to report or make carbon claims, particularly, for carbon neutrality, certain Scope 3 emissions should be reported. However, there is still a degree of flexibility in terms of which Scope 3 emissions are accounted for.

While a GHG inventory approach to calculating emissions may include some Scope 3 emissions, depending on the organisation's discretion, a life cycle approach must include all Scope three emissions contained along the supply chain and within the boundary of the product. More discussion on Scopes was provided in Chap. 11, in examining Australia's Carbon Neutral Certification.

[3] This provides a more accurate representation of the entire emissions attributed to an organisation or product.

Timeframe

Another aspect requiring clarification is the timeframe in which a development will reach carbon neutrality. Buildings are often given a 'lifetime' (i.e. typically 50 years), which forms the basis for the Life Cycle Analysis. This ultimately determines how the emissions that are attributed to the building are spread over the lifetime of that building, and thus how long it will take to achieve carbon neutrality (ASBEC 2011).

Furthermore, timeframes are also needed when determining payback periods, which is particularly significant considering green buildings and developments generally require higher upfront financial expenditure due to the more innovative technologies and designs. It is important to understand the relationship between the higher upfront costs and the long-term savings that occur as a result.

It is likely to be considerably more difficult to establish a timeframe for an entire development compared to a building, due to the different lifetimes associated with different types of infrastructure, for instance buildings, roads, technologies and resource management systems (Sturgis and Roberts 2010). However, it is a critical element to understand and establish, as it will determine the timeframe by which the development can legitimately reach carbon neutrality and the offsets required to attain it. Operational energy is an important part of this timeframe.

In terms of GHG reporting and particularly carbon neutrality claims, the timeframe applied is generally annual, as it is often based on operational emissions. This book therefore recommends a combined life cycle and operational approach to calculating GHG emissions, as a development claiming carbon neutrality is likely to be required to report annually.

Emissions Reduction

Reducing emissions can be done internally or externally. Internal emissions reduction may consist of on-site energy efficiency measures, fuel switching, changing suppliers (i.e. choosing suppliers with the lowest carbon footprint), installing renewable energy on-site and purchasing certified green electricity (UK Department of Communities and Local Government 2009b). External emissions reduction is generally identified as offsetting, which is discussed below.

Internal emissions reduction forms a critical part of the carbon neutral process and is generally required by most schemes to be demonstrated prior to offsetting (City of Sydney 2011; DCCEE 2011a; UK Department of Communities and Local Government 2009a; The Carbon Neutral Company 2012).

Offsets

The term offsets typically refers to the purchase of carbon abatement credits from projects that are outside the operational control of the organisation wishing to reduce their emissions. The Federal Government has defined carbon offsetting as "reductions or removals of greenhouse gases from the atmosphere by sinks, relative to a business-as-usual baseline. Carbon offsets are tradable and often used to negate (or offset) all or part of another entity's emissions" (DCCEE 2012b).

In terms of the built environment and urban precinct development, it is likely that offsets will be required, at the very minimum, to neutralise the emissions associated with the embodied energy within the materials used to create the development. Offsets could also be used to counteract emissions associated with other aspects of the development (e.g. operational emissions, if the developer chose to source the energy required for the precinct from the current fossil fuel-dominated electricity grid). However, as mentioned above, most carbon neutral schemes and standards specify that internal emissions reductions should be pursued as much as possible prior to offsetting, which often means addressing operational emissions.

There are currently no specified limits regarding the amount of offsets that can be purchased compared to on-site emissions reductions, although there are guidelines around which offsets are eligible under certain standards.

It is common within the development industry to classify surplus renewable energy generated on-site and fed into the grid as a carbon offset that can then be used against other emissions sources within the development, such as embodied emissions in materials (Williams 2012). This is often referred to as 'netting' (Sustainability Victoria 2012), and involves balancing the carbon used on-site with sufficient carbon free renewable electricity *produced* on-site and exported to the grid over a one year time period. The development is thus, 'net zero' on an annual basis (Hernandez and Kenny 2009; Williams 2012).

However, this concept is generally not accepted under official carbon neutral certification schemes, which use internationally recognised protocols for offsetting, such as the Australian Government's National Carbon Offset Standard (discussed in Chap. 11). This is because such offsets (i.e. energy related) have until recently not been considered additional forms of abatement,[4] as the energy sector in Australia was a covered sector under the Federal Government's previous Carbon Pricing legislation, and thus any action to reduce emissions from this sector is argued to have occurred anyway due to the mechanism already in place (DCCEE 2012b).

A developer can still use this 'netting' concept as a way of generating their own offsets to achieve carbon neutrality, but they will not be able to achieve this status through the Australian Government's NCOS Carbon Neutral scheme or other schemes that require stringent offsetting standards. It is uncertain whether Australia's new Federal Government policy, the Direct Action Plan, will be able to

[4] As required by the National Carbon Offset Standard.

generate credits from such activities that could be used to offset other emissions from the development as a form of netting.

The development of other kinds of local offsets (i.e. those created within a precinct or community) will still need to meet the 'additionality' criteria defined under the standards to be deemed eligible. Unfortunately, there are few options available for developing offsets from city/community-based activities under NCOS (currently), although there is nothing preventing communities from developing and using these offsets (as long as they provide a clear methodological approach to calculating emission reductions) to claim carbon neutrality without NCOS branding/certification.

References

Australian Sustainable Built Environment Council (ASBEC). (2011). *Defining zero emission buildings. Review and recommendations: Final report.* Retrieved from http://www.asbec.asn.au/files/ASBEC_Zero_Carbon_Definitions_Final_Report_Release_Version_15112011_0.pdf.

Berners-Lee, M., Howard, D. C., Moss, J., Kaivanto, K., & Scott, W. A. (2011). Greenhouse gas footprinting for small businesses—The use of input-output data. *Science of the Total Environment, 409*, 883–891.

Bolwig, S., & Gibbon, P. (2009). Counting carbon in the marketplace: Part 1-overview paper: Report for the OECD. *Trade and Agriculture Directorate; Joint Working Party on Trade and Environment.*

City of Sydney. (2011). *Sustainable sydney 2030: Community strategic plan.* Retrieved from http://cdn.sydney2030.com.au/documents/Community-Strategic-Plan-2011.pdf.

Dhakal, S. (2010). GHG emissions from urbanisation and opportunities for urban carbon mitigation. *Current Opinion in Environmental Sustainability, 2*, 277–283.

Department of Climate Change and Energy Efficiency (DCCEE). (2011a). *Carbon neutral program guidelines: National carbon offset standard.* Version 2.0. Retrieved from http://www.climatechange.gov.au/sites/climatechange/files/documents/03_2013/cnp-guidelines.pdf.

Department of Climate Change and Energy Efficiency (DCCEE). (2011b, July). *National greenhouse gas account factors.* Commonwealth of Australia: Canberra.

Department of Climate Change and Energy Efficiency (DCCEE). (2012b). *National Carbon Offset Standard. Version 2.* Retrieved from http://www.climatechange.gov.au/ncos

East, A. J. (2008). What is a carbon footprint? An overview of definitions and methodologies. *Vegetable industry carbon footprint scoping study.* Sydney: Horticulture Australia Ltd.

Galli, A., Wiedmann, T., Ercin, E., Knoblauch, D., Ewing, B., & Giljum, S. (2011). Integrating ecologial, carbon and water footprint into a "Footprint Family" of indicators: Definition and role in tracking human pressure on the planet. *Ecological Indicators, 16*, 100–112.

Garnaut, R. (2008). *The Garnaut climate change review—final report.* Melbourne: Cambridge University Press.

Hammond, G. (2007). Time to give due weight to the 'carbon footprint' issue. *Nature, 445*(7125), 256.

Hernandez, P., & Kenny, P. (2009). From net energy to zero energy buildings: Defining life cycle zero energy buildings (LC-ZEB). *Energy and Buildings, 42*(6), 815–821.

Hoornweg, D., Sugar, L., & Gomez, C. L. T. (2011). Cities and greenhouse gas emissions: moving forward. *Environment and Urbanization, 23*(1), 207–227.

ICLEI. (2009). *International local government GHG Emissions analysis protocol (IEAP).* Retreived from http://carbonn.org/fileadmin/user_upload/carbonn/Standards/IEAP_October2010_color.pdf.

ICLEI. (2010). Cities in a post-2012 climate policy framework. *ICLEI Global Reports*. Retrieved from http://ccsl.iccip.net/cities_in_a_post_2012_policy_framework.pdf.

International Organisation for Standardisation (ISO). (2006). *Greenhouse gases part 1: Specification with guidance at the organization level for quantification and reporting of greenhouse gas emissions and removals*. International Organisation for Standardisation.

Kennedy, S., & Sgouridis, S. (2011). Rigorous classification and carbon accounting principles for low and zero carbon cities. *Energy Policy*, *39*, 5259–5268.

Matthews, H. S., Hendrickson, C. T., & Weber, C. L., (2008). The importance of carbon footprint estimation boundaries. *Environmental Science and Technology 42*, 5839–5842.

Nässén, J., Holmberga, J., Wadeskog, A., & Nyman, M. (2007). Direct and indirect energy use and carbon emissions in the production phase of buildings: An input-output analysis. *Energy, 32*, 1593–1602.

Rauland, V., & Hall, S. (2012). *South Fremantle senior high school: The journey to carbon neutral*. Retrieved from http://simplycarbon.files.wordpress.com/2012/07/simply-carbon-e-brochure1.pdf.

Rees, W., & Wackernagle, M. (1996). Urban ecological footprints: Why cities cannot be sustainable—and why they are key to sustainability. *Environmental Impact Assessment Review, 16*, 223–248.

Standards Australia. (2006). *Greenhouse gases part 1: Specification with guidance at the organisation level for quantification and reporting of greenhouse gas emissions and removals*. Australian Standards.

Sturgis, S., & Roberts, G. (2010). *Redefining zero: Carbon profiling as a solution to whole life carbon emissions measurement in buildings*. London: RICS Research.

Sustainability Victoria. (2012). *Carbon neutral offices: The pixel building case study. Pamphlet*. Retrieved from http://www.resourcesmart.vic.gov.au/documents/Pixelcasestudy.pdf.

The Carbon Neutral Company. (2012). *The carbon neutral protocol: The global standard for carbon neutral programmes*. Retrieved from http://www.carbonneutral.com/interface/files/aboutus/qualityassurance/TheCarbonNeutralProtocol.pdf.

UK Department of Communities and Local Government. (2009a). *Guidance on carbon neutrality*. London: CLG. Retrieved from http://webarchive.nationalarchives.gov.uk/20110508074721/http://decc.gov.uk/assets/decc/What%20we%20do/A%20low%20carbon%20UK/carbonneutrality/1_20090930090921_e_@@_carbonneutralityguidance.pdf.

UK Department of Communities and Local Government. (2009b). *Zero carbon for new non-domestic buildings: impact assessment*. London: CLG. Retrieved from https://www.gov.uk/government/uploads/system/uploads/attachment_data/file/16153/1391338.pdf.

Wiedmann, T., & Minx, J. (2008). A definition of "Carbon Footprint". *Ecological economic research trends*. NY, USA: Nova Science Publishers.

Williams, J. (2012). *Zero carbon homes: A road map*. New York: Earthscan.

World Resources Institute & World Business Council for Sustainable Development (WRI, and WBCSD). (2004). *The greenhouse gas protocol: A corporate accounting and reporting standard-REVISED EDITION*. Retrieved from http://www.ghgprotocol.org/files/ghgp/public/ghg-protocol-revised.pdf.

Appendix B
Comparison of Precinct-Scale Assessment Schemes

V. Rauland and P. Newman, *Decarbonising Cities*, Green Energy and Technology, DOI 10.1007/978-3-319-15506-7

Table B.1 Comparison of precinct-scale assessment schemes

Tool name	Organisation	Country	Type of tool	Themes/categories targeted	Measurement system	Use of weighting	Phase targeted
BREEAM communities	UK green building council	UK	An international sustainability rating and certification tool	• Climate and energy • Community • Place Shaping • Transport • Resources • Ecology • Business • Buildings	Uses points system (1–100), final score given as a percentage. 51 criteria split between eight categories. Up to 3 credits under each criteria. Not all the categories and credits are mandatory	Yes	Design
LEED neighbourhood development (ND)	US green building council	USA	An international sustainability rating and certification tool	• Smart location and linkage • Neighborhood pattern and design • Green Infrastructure and buildings • Innovation and design process • Regional priority	Points-based system. 53 criteria spread over three primary categories and two additional ones. The number of credits per criteria varies (i.e. some criteria are awarded up to 10 credits). All criteria are mandatory	Yes	Design
CASBEE Urban development (UD)	Green building council	Japan	A national sustainability-rating and certification tool	• Natural environmental quality in urban development; • Service function for the designated area; • Contribution to the local community; • Environmental impact on microclimates, facades and landscape; • Social infrastructure; • Management of the local environment	80 criteria split between 4–6 categories. Each criterion awarded between 1–5 points. Overall score is averaged between categories	Yes	Design

(continued)

Table B.1 (continued)

Tool name	Organisation	Country	Type of tool	Themes/categories targeted	Measurement system	Use of weighting	Phase targeted
One planet communities	BioRegional	International	An international sustainability design and branding tool	• Zero carbon • Zero waste • Sustainable transport • Sustainable materials • Local and sustainable food • Sustainable water • Land use and wildlife • Culture and heritage • Equity and local economy • Health and happiness	Assesses unique targets set by individual developments to ensure they comply with the 10 principles. Requires annual reviews to ensure action plan is being delivered and targets met	No	Design and performance (annual review)
Green star communities	Green building council of Australia	Australia	A national sustainability rating and certification tool	• Governance • Design • Liveability • Economic prosperity • Environment • Innovation	Uses points system (1–100) to give overall rating in stars (4–6 stars). 38 credits available over 6 categories	Yes	Design
Enviro-development	Urban development institute of Australia (UDIA)	Australia	A national sustainability branding and certification system	• Ecosystems • Waste • Energy • Materials • Water • Community	Points-based. Uses leaves. Uses technical standards and requires per cent improvement. Can certify a single area or up to all six	No	Design. Also offers ongoing certification based on performance rating tools

(continued)

Table B.1 (continued)

Tool name	Organisation	Country	Type of tool	Themes/categories targeted	Measurement system	Use of weighting	Phase targeted
PRECINX	Landcom (NSW Government Land Development Agency)	Australia	Quantitative decision-making tool for urban development. No certification	• On-site energy • Embodied CO_2 • Potable water • Transport • Housing diversity • Storm water	Uses quantifiable numbers to measure predicted performance i.e. tonnes of GHG's, mL of water, vehicle kilometres traveled (VKT), affordability	Yes	Design
eTool	eTool	Australia	Life cycle assessment (LCA) software to aid low carbon design. Certification possible	• Carbon associated with life cycle of buildings, developments and infrastructure projects	Quantitative Life Cycle Assessment of carbon within a development	No	Design and performance

Appendix C
ISO Definitions

The ISO Definitions of Accreditation and Certification of GHG Assertions

The International Organisation for Standardisation (ISO) was established in 1947 "to facilitate the international coordination and unification of industrial standards" (ISO 2013).[5] Over the past 65 years it has developed over 19,000 standards spanning a range of fields and industries, bringing greater consistency and credibility to business practices worldwide.

In relation to carbon and GHG accounting, ISO has developed a range of standards, including the ISO 14064 series that provides guidance on calculating GHG emissions, as well as on the verification of GHG claims and assertions. Another more general standard relevant to the carbon industry relates to conformity assessment. The standard, known as ISO/IEC 17000:2004 Conformity Assessment—Vocabulary and General Principles (ISO 2004), helps to identify credible bodies that carry out GHG audits and providing certification. These bodies ensure stringent adherence to the guidelines and methods outlined in the ISO 14064 series. They are referred to as 'Conformity Assessment Bodies' or CABs and their accreditation has been described as "third party attestation related to a conformity assessment body conveying formal demonstration of its competence to carry out specific conformity assessment tasks" (ISO 2004). These organisations or CABs are accredited based on the international standard ISO 14065:2007 *Greenhouse gases—Requirements for greenhouse gas validation and verification bodies for use in accreditation or other forms of recognition.*

[5] http://www.iso.org/iso/about/the_iso_story/iso_story_founding.htm.

V. Rauland and P. Newman, *Decarbonising Cities*, Green Energy and Technology, DOI 10.1007/978-3-319-15506-7

ISO 14064 Greenhouse Gas Series and the GHG Protocol

The ISO 14064 Greenhouse Gas series consists of three documents. They include:

1. ISO 14064-1:2006. Greenhouse gases—Part 1: Specification with guidance at the organisation level for quantification and reporting of greenhouse gas emissions and removals;
2. ISO 14064-2:2006. Greenhouse gases—Part 2: Specification with guidance at the project level for quantification, monitoring and reporting of greenhouse gas emissions reductions or removal enhancement; and
3. ISO 14064-3:2006. Greenhouse gases—Part 3: Specification with guidance for the validation and verification of greenhouse gas assertions.

The ISO series of standards are based predominantly on the Greenhouse Gas Protocol documents, which were produced as part of the Greenhouse Gas Protocol Initiative. The initiative convened in 1998 and was a collaboration between the World Business Council for Sustainable Development (WBCSD) and the World Resources Institute (WRI). Three primary documents were produced as part of the initiative and include:

1. The Greenhouse Gas Protocol—A Corporate Accounting and Reporting Standard;
2. The Greenhouse Gas Protocol—The GHG Protocol for Project Accounting; and
3. GHG Protocol—Corporate Value Chain (Scope 3) Accounting and Reporting Standard (2011).

The three documents were developed in 2001, 2004 and 2011 respectively. They provide detailed guidance on GHG accounting and reporting for projects and organisations and form the basis for most GHG accounting schemes worldwide. The ISO 14064 series was produced through a partnership between WBCSD, WRI and ISO.

In addition to the ISO, Australia has its own organisation for standardisation called Standards Australia. This organisation originated as an association in 1922, before the creation of the ISO. However, it became a founding member of the ISO in 1947 to ensure that the standards developed would be internationally aligned.[6] As such, these internationally aligned standards are labelled 'AS ISO' standards. The Australian version of the abovementioned standards is thus 'AS ISO 14064:2006 Greenhouse Gases' series.

[6] See http://www.standards.org.au.

References

International Organisation for Standardisation (ISO). (2004). ISO/IEC 17000:2004 *Conformity Assessment—Vocabulary And General Principles*. Retrieved from http://www.iso.org/iso/catalogue_detail.htm?csnumber=29316.

International Organisation for Standardisation (ISO). (2013). Standards. *Certification* Retrieved April 24, 2013, from http://www.iso.org/iso/home/standards/certification.htm.

Appendix D
Comparisons of Carbon Neutral Certification Schemes

V. Rauland and P. Newman, *Decarbonising Cities*, Green Energy and Technology, DOI 10.1007/978-3-319-15506-7

Table D.1 Comparisons of carbon neutral certification schemes

Company	Name of certification/logo	Est.	Country of origin	Current market	Third party verification required	Type of company/org	ISO 14065 accreditation	Supply offsets
carboNZero	Carbon Zero	2001	New Zealand	International	Yes	Government partnership	Yes	Yes
	CEMARS				No			
Carbon-neutral	CarbonNeutral® Global Standard	1997	UK	International	Yes	Private, for profit	No	Yes
British standards institution	PAS 2060:2010—Specification for the demonstration of carbon neutrality	2010	UK	UK	Yes but no follow up	Government	N/A	No
Carbon neutral	None/Can certify under a range of standards	2001	Australia	Australia	Depends on standard chosen	Private, Not for profit	No	Yes
Climate friendly	Climate friendly business/event taking action with climate Friendly	Not listed	Australia	International	No	Private, for profit	No	Yes
Carbon reduction institute	$NoCO_2$	2006	Australia	Australia	No	Private, for profit	No	Yes
	$LowCO_2$							
	Carbon neutral							
ICLEI (Carbon Neutral Framework)	None	1990	Canada	International	Yes	Association	N/A	No
		2008						
Low carbon Australia	NCOS carbon neutral certification	2010	Australia	Australia	Yes	Government	No	No

Appendix E
White Certificate Schemes

Table E.1 White certificate schemes

Country	Start date	Name of scheme	Summary
Australia (South Australia)	2009	Retailer energy efficiency scheme (REES)	• Energy retailers with more than 5,000 residential customers must surrender sufficient certificates • Targets residential sector and small businesses • Certificates are generated through pre-approved energy efficiency activities
Australia (NSW)	2009	Energy savings scheme (ESS)	• Liable energy retailers must obtain and surrender sufficient certificates to meet their target • Accredited certificate providers generate certificates from recognised energy saving activities (RESA's) • Targets businesses and households
Australia (Victoria)	2009	Victorian energy efficiency target (VEET) scheme	• Energy retailers who are liable under the scheme must purchase and surrender a specified number of victorian energy efficiency certificates (VEEC's) annually • Certificates are generated from prescribed (energy efficiency) activities by accredited entities • Activities can be targeted at both residential and commercial sectors

(continued)

V. Rauland and P. Newman, *Decarbonising Cities*, Green Energy and Technology, DOI 10.1007/978-3-319-15506-7

Table E.1 (continued)

Country	Start date	Name of scheme	Summary
Australia (ACT)	2013	Energy efficiency improvement scheme (EEIS)	• The EEIS is underpinned by the ACT's territory-wide energy target • Energy retailers need to provide sufficient abatement factors (similar to certificates) from eligible activities to meet their individual obligation, though certificates are not created or traded • Targets residential or commercial customers
India	2011	Perform, achieve and trade system (PAT)	• Designated consumers (energy intensive industries and facilities) that are responsible for 54 % of emissions • Purchase tradable energy savings certificates
United Kingdom	2010	Carbon reduction commitment (CRC) energy efficiency scheme	• Energy suppliers with over 50,000 customers (residential) • Large public and private sector (i.e. supermarkets, banks, government departments)
United Kingdom	2010	CRC energy efficiency scheme	• Tackles emissions not covered under current ETS • Cap and trade scheme targeting non energy-intensive industries • Organisations whose annual electricity use in 2008 exceeded 6M wH required to report emissions
Italy	2005	Energy efficiency certificates (TEE: Titoli Efficienza Energetica)	• Distributors of electricity and gas with more than 100,000 customers are obliged to deliver a certain number of white certificates per year
France	2006	French white certificates	• Energy suppliers with minimum 400 GWh/year sales and all suppliers of residential heating oil
Denmark	2006	Energy efficiency obligation (Looking to change to trading scheme)	• Obligations are put on electricity, gas, district heating and oil companies

MIX
Papier aus verantwortungsvollen Quellen
Paper from responsible sources
FSC® C105338

If you have any concerns about our products,
you can contact us on
ProductSafety@springernature.com

In case Publisher is established outside the EU,
the EU authorized representative is:
Springer Nature Customer Service Center GmbH
Europaplatz 3, 69115 Heidelberg, Germany

Printed by Libri Plureos GmbH
in Hamburg, Germany